T0205470

Textual and Visual Information Retrieval using Query Refinement and Pattern Analysis

S. G. Shaila · A. Vadivel

Textual and Visual Information Retrieval using Query Refinement and Pattern Analysis

Springer

S. G. Shaila
Department of Computer Science
  and Engineering
Dayananda Sagar University
Bangalore, India

A. Vadivel
Department of Computer Science
  and Engineering
SRM University AP
Amaravati, Andhra Pradesh, India

ISBN 978-981-13-4791-7   ISBN 978-981-13-2559-5   (eBook)
https://doi.org/10.1007/978-981-13-2559-5

This Springer imprint is published by the registered company Springer Nature Singapore Pte Ltd.
The registered company address is: 152 Beach Road, #21-01/04 Gateway East, Singapore 189721, Singapore

This book is dedicated to my guide, who is the co-author of the book, my parents and my family. Their encouragement and understanding helped me to complete this work. I thank my husband and children, Vinu and Shamitha, for their love and support to bring this book as a reality.

Foreword

I am extremely delighted to write the foreword for *Textual and Visual Information Retrieval using Query Refinement and Pattern Analysis*. The authors, Dr. S. G. Shaila and Dr. A. Vadivel, have disseminated their knowledge on information retrieval through this book. The content of the book is a very good educational and valuable resource for researchers in the domain of information retrieval.

The book includes various chapters on deep Web crawler, event pattern retrieval, Thesaurus generation and query expansion, CBIR applications and indexing and encoding to cover the whole concept of information retrieval. Each chapter contains the theoretical information with experimental results and is intended for information retrieval researchers. The layout of each chapter includes a table of contents, introduction, material content and experimental results with analysis and interpretation.

This edition of the book reflects new guidelines that have evolved in information retrieval in terms of text- and content-based information retrieval schemes. The indexing and encoding mechanism of the low-level feature vector is also presented with results and analysis.

It is my hope and expectation that this book will provide an effective learning experience and referenced resources for all information retrieval researchers, leading to advanced research.

Bangalore, Karnataka, India

Dr. M. K. Banga
Chairman
Department of Computer
Science Engineering

Dean (Research)
School of Engineering
Dayananda Sagar University

Preface

Multimedia information retrieval from the distributed environment is an important research problem. It requires an architecture specification for handling various issues such as techniques to crawl information from WWW, user query prediction and refinement mechanisms, text and image feature extraction, indexing and encoding, similarity measure. In this book, research issues related to all the above-mentioned problems are discussed and suitable techniques are presented in various chapters.

Both the text and images are presented in web documents. In a comprehensive retrieval mechanism, text-based information retrieval (TBIR) plays an important role. In Chap. 1, the text-based retrieval is used for retrieving relevant documents from the Internet by using a suitable crawler with the capability to crawl deep and surface web. The functional dependency of core and allied fields in HTML FORM is identified for generating rules using SVM classifier. The presented crawler fetches a large number of documents while using the values in most preferable class. This architecture has a higher coverage rate and reduces fetching time.

In recent times, information classification is very important for text-based information retrieval. In Chap. 2, the classification based on events is presented and also the event Corpus is discussed, which is important for many real-time applications. Event patterns are identified and extracted at the sentence level using term features. The terms that trigger events along with the sentences are extracted from web documents. The sentence structures are analysed using POS tags. A hierarchal sentence classification model is presented by considering specific term features of the sentence, and the rules are derived along with fuzzy rules to get the importance of all term features of the sentences. The performance of the method is evaluated for 'Crime' d 'Die' and found that the performance of this approach is encouraging.

In general, the retrieval system depends on the user query to retrieve the web documents. The user-defined queries should have sufficient relevant terms, since the retrieval set depends on the queries. The query refinement through query expansion mechanism plays an important role. In Chap. 3, the N-gram Thesaurus construction mechanism for query expansion is presented. The HTML TAGs in web documents are considered and their syntactical context is understood. Based on the significance of the TAGs in designing the web pages, suitable weight is assigned for TAGs. The term weight is calculated using corresponding TAG weight and frequency of the term. The terms along with the TAG information are updated into an inverted index. The N-grams are generated using the term and term weights in the document and updated as N-grams in the Thesaurus. During the query session, the term is expanded based on the content in the Thesaurus and suggested to the user. It is found that while the selected query is submitted to the retrieval system, the retrieval set consists of a large number of relevant documents.

In Chap. 4, the issues related to content-based image retrieval (CBIR) are presented. The chapter presents a histogram based on human colour visual perception by extracting the low-level features. For each pixel, the true colour and grey colour proportion are calculated using a suitable weight function. During histogram construction, the hue and intensity values are iteratively distributed to the neighbouring bins. The NBS distance is calculated between the reference bin and their adjacent bins. The NBS distance provides the overlapping proportion of the colour from the reference bin to its adjacent bins, and accordingly, the weight is updated. The distribution makes it possible to extract the background colour information effectively along with the foreground information. The low-level features of all the images in the database are extracted and stored in a feature database, and the relevant images are retrieved based on the rank. The Manhattan distance is used as a similarity measure, and the performance of the histogram is evaluated on Coral benchmark dataset.

In Chap. 5, the issues of indexing and encoding of low-level features and a similarity measure are presented. In CBIR system, the low-level features are stored along with the images and require a large number of storage space along with increased search and retrieval time. The search time increases linearly with the database size, which reduces the retrieval performance. The colour histograms of images are considered as low-level features. The bin values are analysed to understand their contribution representing image colour. The trivial bins are truncated and removed, and only important bins are considered to have histograms with lesser number of bins. The coding scheme used GR coding algorithm, and the quotient and remainder code parts are evaluated. Since there is variation between the number of bins in the query and database histogram, *BOSM* is used as a similarity measure. The performance of all the schemes is evaluated in an image retrieval system. The retrieval time, number of bits needed for histogram construction and precision of retrieval are evaluated using benchmark datasets, and the performance of the presented approach is encouraging.

Finally, as a whole, the book presents various important issues in information retrieval research filed and will be very much useful for the postgraduates and researchers working in information retrieval problems.

Bangalore, India   S. G. Shaila
Amaravati, India   A. Vadivel

Acknowledgements

First and foremost, we thank the Almighty for giving the wisdom, health, environment and people to complete this book.

We express our sincere gratitude to Dr. Hemachandra Sagar and Premachandra Sagar, Chancellor and Pro-Chancellor, Dayananda Sagar University, Bangalore; Dr. A. N. N. Murthy, Vice Chancellor, Dayananda Sagar University, Bangalore; Prof. Janardhan, Pro-Vice Chancellor, Dayananda Sagar University, Bangalore; Dr. Puttamadappa C., Registrar, Dayananda Sagar University, Bangalore; Dr. Srinivas A., Dean, School of Engineering, Dayananda Sagar University, Bangalore; Dr. M. K. Banga, Chairman, Department of CSE, and Dean Research, Dayananda Sagar University, Bangalore, for providing an opportunity and motivation to write this book.

We express our sincere gratitude to Dr. P. Sathyanarayanan, President, SRM University, Amaravati, AP; Prof. Jamshed Bharucha, Vice Chancellor, SRM University, Amaravati, AP; Prof. D. Narayana Rao, Pro-Vice Chancellor, SRM University, Amaravati, AP; Dr. D. Gunasekaran, Registrar, SRM University, Amaravati, AP, for providing an opportunity and motivation to write this book.

We would like to express our sincere thanks to our parents, spouse, children and faculty colleagues for their support, love and affection. Their inspiration gave us the strength and support to finish the book.

Dr. S. G. Shaila
Dr. A. Vadivel

Contents

About the Authors

Dr. S. G. Shaila is an Associate Professor in the Department of Computer Science and Engineering in Dayananda Sagar University, Bangalore, Karnataka. She earned her Ph.D. in computer science from the National Institute of Technology, Tiruchirappalli, Tamil Nadu, for her thesis on multimedia information retrieval in distributed systems. She brings with her years of teaching and research experience. She worked for the DST project, Govt of India and Indo US based projects as Research Fellow. Her main areas of interest are information retrieval, image processing, cognitive science and pattern recognition. She published 20 international journals and confernce proceedings.

Dr. A. Vadivel received his master's in science from the National Institute of Technology, Tiruchirappalli (NITT), Tamil Nadu, before completing a master's in Technology (M.Tech.) and Ph.D. at the Indian Institute of Technology (IIT), Kharagpur, India. He has 12 years of technical experience as a network and instrumentation engineer at the IIT Kharagpur and 12 years of teaching experience at Bharathidasan University and NITT. Currently, he is working as Associate Professor at SRM University, Amaravati, AP. He has published papers in more than 135 international journals and conference proceedings. His research areas are content-based image and video retrieval, multimedia information retrieval from distributed environments, medical image analysis, object tracking in motion video and cognitive science. He received the Young Scientist Award from the Department of Science and Technology, Government of India, in 2007; the Indo-US Research Fellow Award from the Indo-US Science and Technology Forum in 2008; and the Obama-Singh Knowledge Initiative Award in 2013.

Abbreviations

ACE	Automatic Content Extraction
ALNES	Active Long Negative Emotional Sentence
ALNIES	Active Long Negative Intensified Emotional Sentence
ALPES	Active Long Positive Emotional Sentence
ALPIES	Active Long Positive Intensified Emotional Sentence
ANFIS	Artificial neuro-fuzzy inference system
ANN	Artificial neural network
ASNES	Active Short Negative Emotional Sentence
ASNIES	Active Short Negative Intensified Emotional Sentence
ASPES	Active Short Positive Emotional Sentence
ASPIES	Active Short Positive Intensified Emotional Sentence
Bo1	Bose–Einstein statistics model
BoCo	Bose–Einstein statistics co-occurrence model
BOSM	Bin overlapped similarity measure
CART	Classification and regression tool
CBIR	Content-based image retrieval
CF	Core field
Co	Co-occurrence
CRF	Conditional random fields
DCT	Discrete cosine transform
DOM	Document object model
EMD	Earth mover's distance
ET	Emotional triggered
FGC	Form Graph Clustering
FPN	Fuzzy Petri-Nets
GR	Golomb–Rice
HCPH	Human colour perception histogram
HiWE	Hidden Web Exposer
HQAOS	High-Qualified Active Objective Sentence
HQASS	High-Qualified Active Subjective Sentence

HQPOS	High-Qualified Passive Objective Sentence
HQPSS	High-Qualified Passive Subjective Sentence
HSV	Hue Saturation Value
HTML	Hyper-Text Markup Language
IBC	Iraq Body Count
InPS	Internal Property Set
IPS	Input Property Set
IR	Information retrieval
IRM	Integrated Region Matching
IWED	Integrated Web Event Detector
KDB	K-Dimensional B-tree
KLD	Kullback–Liebler divergence
KLDCo	Kullback–Liebler divergence co-occurrence model
LDA	*Latent Dirichlet allocation*
LP	Least Preferable
LVS	Label value set
MAP	Mean average precision
MCCM	Colour-based co-occurrence matrix scheme
ME	Mutually Exclusive
MEP	Minimum Executable Pattern
MHCPH	Modified human colour perception histogram
MP	Most Preferable
MRR	Mean reciprocal recall
MUC	Message Understanding Conference
NBS	National Bureau of Standards
NIST	National Institute of Standards and Technology
NLP	Natural language processing
NQAOS	Non-Qualified Active Objective Sentence
NQASS	Non-Qualified Active Subjective Sentence
NQPOS	Non-Qualified Passive Objective Sentence
NQPSS	Non-Qualified Passive Subjective Sentence
OPS	Output Property Set
PHOTO	Pyramid histogram of topics
PIW	Publicly Indexable Web
PLNES	Passive Long Negative Emotional Sentence
PLNIES	Passive Long Negative Intensified Emotional Sentence
PLPES	Passive Long Positive Emotional Sentence
PLPIES	Passive Long Positive Intensified Emotional Sentence
POS	Part of speech
PQ	Product quantization
PSNES	Passive Short Negative Emotional Sentence
PSNIES	Passive Short Negative Intensified Emotional Sentence
PSPES	Passive Short Positive Emotional Sentence
PSPIES	Passive Short Positive Intensified Emotional Sentence
QA	Question Answering

QAOS	Qualified Active Objective Sentence
QASS	Qualified Active Subjective Sentence
QPOS	Qualified Passive Objective Sentence
QPSS	Qualified Passive Subjective Sentence
RED	Retrospective new Event Detection
RS	Rule set
SCM	Sentence classification model
SOAP	Simple object access protocol
SQAOS	Semi-Qualified Active Objective Sentence
SQASS	Semi-Qualified Active Subjective Sentence
SQPOS	Semi-Qualified Passive Objective Sentence
SQPSS	Semi-Qualified Passive Subjective Sentence
SVM	Support vector machine
SWC	Surface web crawlers
SWR	Semantic Web Retrieval
TBIR	Text-based information retrieval
TDT	Topic Detection and Tracking
TS	Text Summarization
TSN	Term semantic network
TTW	TAG term weight
UMLS	Unified Medical Language System
URL	Uniform Resource Locator
ViDE	Vision-based data extraction
WaC	Web as Corpus
WSDL	Web Services Description Language
WWW	World Wide Web
XML	Extensible Markup Language

List of Figures

List of Tables

Chapter 1
Intelligent Rule-Based Deep Web Crawler

1.1 Introduction to Crawler

In earlier days, the number of documents in WWW was relatively less, and thus, managing and fetching them for processing is easy. The crawler is used as a tool by most of the search engine systems for fetching these static-natured documents. However, WWW has grown fast with thousands to million number of web pages. The content of HTML is also altered, uploaded by authors often. This makes the retrieval task complex and difficult to achieve good precision of retrieval. The retrieval result is also influenced by the user's query, and search engine systems process the user query suitably for retrieving relevant results. The web pages are crawled by the crawler periodically and are stored in the repositories for continuously updating the content. In recent scenario, the content of WWW is hidden in the backend and referred to as deep web. The web applications use these hidden information for dynamically creating web pages, which is the most frequently retrieved information by dynamic web-based applications. These information is invariably not available to the all well-known surface web crawlers, since because, hidden nature of database information. Most of the web-based applications use searchable data sources, and this kind of information is referred to as deep web. This content is dynamically retrieved by the users. All these database systems provide tools for performing database-related analysis and search process.

In general, the crawler fetches documents from Publicly Indexable Web (PIW) (Alvarez et al. 2007). The PIW consists of set of web pages with interconnected hypertext links. The pages with FORMs where user provides username and password for authorization are not visible these kinds of crawler. The advanced web-based applications demands more web databases are created and used. The data stored in such as database can be accessed through search interfaces only. In fact, there are huge number of deep web databases, and query interfaces are hidden with very high increasing rate (Ajoudanian and Jazi 2009). The number of hidden pages is very high in number compared to PIW, which shows that large number of information

© Springer Nature Singapore Pte Ltd. 2018
S. G. Shaila and A. Vadivel, *Textual and Visual Information Retrieval using Query Refinement and Pattern Analysis*, https://doi.org/10.1007/978-981-13-2559-5_1

is hidden and directly is not accessible by the surface web crawlers. The difference between the surface and deep web crawlers is the logical challenge in fetching the information from deep and surface web. The Googlebot (Brooks 2004) and Yahoobot (Gianvecchio et al. 2008) are well-known crawler and indexers. These applications are not capable of accessing the web databases. The reason is that the information from the database will be retrieved after performing computation. As a result, it is important to understand the query interface, handling the technical challenges such as interaction with them, locating, accessing and indexing. Also, the web crawler has to traverse the web automatically, fill the FORMs intelligently, store the fetched data effectively in local repositories and manage all these tasks.

It is understood from the above discussion that an intelligent deep web crawler is need of the hour. The crawler should essentially interact with the FORMs in HTML pages automatically to fill the fields effectively without human intervention. While filling the FORMs, the combinations of FORM element and FORM value have to be predicted. This can be made possible by understanding the FORM with prior knowledge. In this chapter, architecture of a web crawler is presented, where the FORMs in HTML pages are filled by the following rules. The rules are derived for Mutually Exclusive, Most Preferable and Least Preferable classes. In addition, all possible combinations of values are tried, and the crawler efficiency is improved. The rest of the chapter is organized as follows. The similar work is reviewed in Sect. 1.2, and intelligent crawler is presented in Sect. 1.3. In Sect. 1.4, the experimental results are presented and the chapter is concluded in the last section of the chapter.

1.2   Reviews on Web Crawlers

The design and development of crawlers is in steady growth with a large number of design and specification from acadamia and industry. The basic crawler along with heuristics is proposed for increasing the efficiency. Chakrabarti et al. (1999) have proposed a general approach, which uses link structure of the web for analysis. Renni and McCallum (1999) have proposed a machine learning approach for fetching the documents. In recent times, the deep web is considered as potential research interest (Chang et al. 2004). Raghavan and Arasu et al (2001) have developed a web crawler for interacting with hidden web data through web search interfaces. Hidden Web Exposer (HiWE) is a deep web crawler to interact with FORM, and the FORMs are filled using a customized database, which use the layout-based information extraction approach. However, this approach fails due its FORM-filling strategy based on simple dependency. The source-biased probing techniques are used to facilitate interaction with the target database to find its relevancy (Caverlee et al. 2006). The relevancy is calculated using relevance metrics for evaluating interesting content in the hidden web data source. One of the limitations of his approach is that it relies on several key junctures, errors in relevance evaluation and difficult level in identifying relationship.

While processing large number web documents, it is difficult to identify the relevance of deep web content with the query. Zhao et al. (2008) have crawled substantial number of web pages from WWW to structure the sources of the web pages for providing effective query interface for suitable results (Zhao et al. 2008). Organizing structured deep web sources for various domains of web applications is a challenging task. A Form Graph Clustering (FGC) approach has been proposed to classify the content of deep web resources with the help of FORM link graph by suitably applying the fuzzy partition technique on web FORM. This method is suitable for single-domain deep web resources. The schema matching is continuously found using probability of schema. However, there is a higher chance that valuable data are ignored while calculating probability of schema. Ntoulas et al. (2008) have proposed a crawler to discover and download deep web content autonomously. This method generates queries automatically for handling query interface. FORMs are filled using the set covering problem. For each successful query, the number of matching pages is calculated. The list of keywords is randomly chosen as query list using query generation policy. This is using frequency of occurrence of the keyword in generic text collection and content of the web pages fetched from hidden website. As a result, each and every page requires more time to download and storage space. Lie et al. (2011) have proposed Minimum Executable Pattern (MEP), where the query FORM uses minimal combination of elements for performing query (Liu et al. 2011). It works based on MEP generation and MEP-based deep web adaptive crawling approach. The optimal queries are generated by parsing the FORMs and partitioning ME patterns. One of the drawbacks of this approach is that the number of documents fetched by this approach is very low.

Wei et al. (2010) have developed vision-based data extraction (ViDE), which has used the visual content of Web documents. The similarity in visual content among the web documents is analysed for extracting information. However, the ViDE performs good only when there is visual similarity between web content and otherwise fails. Kayed and Chang (2010) have proposed Fiva-Tech where the web data is extracted in page level (Kayed and Chang 2010). The web pages are matched with each other based on a template. Tree alignment method along with mining technique is applied for extracting the data. The distributed object model tree is constructed to understand the pattern of HTML documents. It is found that tree construction time is large and in case if immediate web page differs from the previous one, matching may not be accurate. For every FORM-based query, its distribution in the corpus is used to predict the property of retrieved document (Ntoulas et al. 2005). One of the major drawbacks of this method is that attribute value set and the distributions of queries are not found to be adequate. Wu et al. (2006) have modelled the structured web information as a distinct attribute-value graph to crawl the hidden web information (Wu et al. 2006). However, query of all the nodes require to be inserted in the graph and the cost of the process is high. Madhavan et al. (2008) of Google Corporation have developed a crawler to surf the deep web, where the search space is navigated with possible input combinations for query for identifying only the suitable URLs to include in the search index mechanism. However, the efficiency of the model is not taken into consideration. The query having high frequency of occurrence is selected for seed for crawling the web document (Barbosa and Freire 2004). However, it is not always guaranteed that more new pages are fetched with high-frequency terms.

The domain specification has been used as a parameter for accessing the hidden web pages (Alvarez et al. 2007). Several heuristics based on visual distance and text similarity measures are used. The bounded fields may not have any globally associated text in the Web FORMs. This approach assumes that every FORM field has an association with a text to show the function of the field. It is observed that most of the FORMs may not have association and explanation information of the text label. Yongquan and Qingzhong (2012) have sampled web document, and training set is constructed for selecting multiple features. Suitable query strings are selected in each round of crawling phase till the termination condition is reached. It is noticed that a good coverage rate is achieved, but a large number of duplicate documents are fetched with higher processing time. In addition, Barbosa and Freire (2007) have proposed a learning approach and found that learning iteration and the sample path are high.

It is imperative for the above discussions that fetching deep web data is an important task. Developing a deep web crawler for all the domain web applications is a difficult task. The processing time is bottleneck for the most of web crawler. The web FORMs are filled with all the possible values, and the combination of values is huge. As a result, crawler is ineffective to fetch the web documents from deep web. In addition, none of the approach has indexer included in the architecture for effectively utilizing the ideal time of the crawler. Thus, a deep web crawler is required to crawl the surface web, deep web and indexer. Suitable rule sets are required to understand the FORM values for most of the domain-specific applications. In this chapter, an architecture specification of a crawler is proposed for retrieving surface and deep web documents with prior knowledge of the domain using the rules. The relationship between FORM values are found by the rules, and the values to FORMs are filled effectively to achieve higher coverage rate.

1.3   Deep and Surface Web Crawler

In this section, the details about the crawler are presented and explained. Each web application consists of FORMs with input elements along with values. Each FORM element has a label and name relationship. As a result, the FORM is written as $F = (\{e_1, e_2, e_3, \ldots, ei\}, S, M)$ where e_i is the number of FORM elements, S is information to be submitted. The URL of the FORM, web address and number of pages connected to the FORM are denoted by M. L is label, and $V = \{v_1, v_2, v_3, \ldots, v_i\}$ are set of values. The labels are assigned suitable values. Thus, each v_i is a value that is assigned to appropriate element e_i if label e is equal to L. For a given domain, suitable and probable values are available in the FORMs. The text boxes in the FORMs are free form inputs as these kinds of input elements are entered by the user without any constraint. Similarly, input in the form of descriptive text is also considered as free form such that the users understand the meaning of the element. The FOMRs of a web page for a domain-specific web page, Label-Value-Set (LVS) table is extracted for further processing.

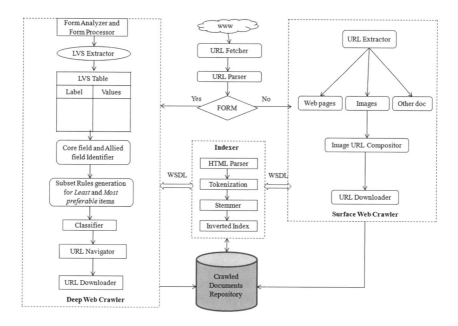

Fig. 1.1 Block diagram of deep web crawler

The crawler has various blocks with rule specification for fetching deep web content and is shown in Fig. 1.1. The figure contains three functional units such as deep web crawler, surface web crawler along with indexer; there are three blocks such as surface web, deep web and indexer. The conventional HTML pages are fetched by the surface web crawler. The web page is located and fetched by the URL fetcher from the WWW, and the URL links are verified by the URL parser to avoid the dead link. The FORMs in a HTML document are located by understanding the TAG values. While the HTML page contains <FORM> HTML TAG, the deep web crawler fetches the document as shown in Fig. 1.1. The FORM elements are filled, and it is submitted as query to the web applications. The response to the query is created dynamically based on the values to FORM elements, which is the content of the deep Web. The function of the FORM processor is self-explanatory by which the form filling, submission, and retrieving dynamic web pages are handled.

Each FORM contains a large number of input fields. All or some of the fields are filled by the users, and the LVS table is constructed. However, the allied and core fields combination estimated by the rules may vary drastically for different web applications.

1.4   Estimating the Core and Allied Fields

Given two sets, $\{i = 1, 2, \ldots n\}$ and $\{j = 1, 2, \ldots m\}$. The functional dependency between given sets is referred to as constraint on the attributes belong to the sets. A set of attributes $AF1_i \in R_1$ in relation R_1 will functionally determine another attribute $AF2_j \in R_2$, if and only if, $AF1_i \to AF2_j$. Thus, $AF1_i$ is the determinant attribute and $AF2_j$ is dependent attribute. As a result, if value of $AF1_i$ is found, the value of $AF2_j$ is estimated approximately. The functional dependency between two attributes are such that one attribute depends on the other. Given that $AF1_i$, $AF2_j$, and CF_l where $\{l = 1, 2, \ldots k\}$ are sets of attributes in a relation R, the axiom of transitivity is used to determine the properties of functional dependencies as given below.

Axiom of transitivity: If $AF1_i \to AF2_j$ and $AF2_j \to CF_l$, then $AF1_i \to CF_l$

The above axiom is applied for deriving the functional dependency between the core and allied fields. The $(AF1_i)$ is functionally dependent on $(AF2_j)$; i.e., $AF1_i \to AF2_j$ and $(AF2_j)$ is functionally dependent on (CF_l), i.e., $AF2_j \to CF_l$, which is shown in Fig. 1.2. In this work, the Core field (CF_l) is constant and $AF1_j$ functionally dependent on CF_l attributes. The rules are constructed in the form of if-then-else construct for the core and allied filed combinations. Various combinations of $AF1_i$, $AF2_j$ and CF_l and classes such as Most Preferable (*MP*), Least Preferable (*LP*) and Mutually Exclusive (*ME*) found. The attribute values of *MP* class retrieve large number of documents from the hidden web. In contrast, the attribute values belonging to LP class retrieve lesser number of documents. The attribute belonging to *ME* class is logically invalid and as the combination will not retrieve any of the documents from the Web.

For a given application domain, CF_l is fixed, and $AF1_i$ as well as $AF2_j$ are suitably chosen. Based on experiments, it is noticed that for a combination of CF_l, $AF1_i$ and $AF2_j$ the rules belonging to most preferable class retrieve large number of documents compared to the rules belonging to least preferable class, which is shown in Fig. 1.3. Here, the value of CF_l and $AF2_j$ is fixed, and the value of $AF1_i$ is substituted. This is represented in Eqs. (1.2–1.4). These equations are derived by analysing the results manually for a domain to suitably select the combination of allied and core fields.

$$\left.\begin{array}{l} AF1_i \to AF2_j \\ AF2_j \to CF_l \end{array}\right\} \Rightarrow AF1_i \to CF_l \tag{1.1}$$

Equations (1.2–1.4) show the relationship between allied and core fields for *MP* and *LP* classes.

$$AF1_{i(y)MP} = (AF2_{j(x)} - 1) + CF_l \tag{1.2}$$

$$AF1_{i(y)LPT} = (AF2_{j(x)} + 1) - 1 + CF_l \tag{1.3}$$

$$AF1_{i(y)LPD} = (AF2_{j(x)} - 1) - 1 + CF_l \tag{1.4}$$

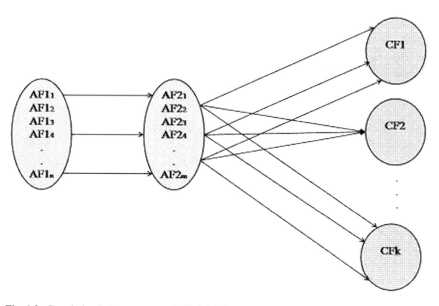

Fig. 1.2   Co-relation between core and allied fields

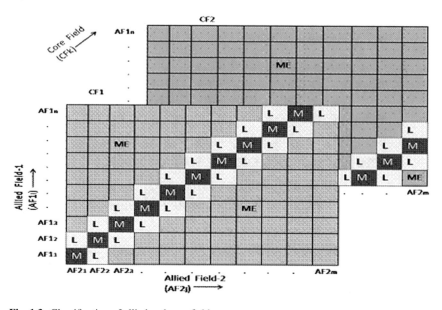

Fig. 1.3   Classification of allied and core fields

The SVM classifier is used for classifying Most preferable and Least Preferable classes by correlating the allied and core fields. The output space of the classifier is binary to denote the *MP* and *LP* classes. The classification scheme for real estate website is presented in Sect. 1.8.1.

1.5   Classification of Most and Least Preferred Classes

The interpretations of the rules are presented below and can be interpreted as follows. Say, for any CF_l the first rule is interpreted as the combination of $AF2_1$ and $AF1_1$, which is Most Preferable and $AF2_1$ and $AF1_2$ is Least Preferable combinations. In general, for $AF2_m$ and $AF1_n$ combination, there is a *MP* class, and next units $AF1_n$ are the *LP* class. While there is *ME* class, the *MPs* and *LPs* do not exist, and in a similar way, the rule is interpreted for various rules. Using these rules, the FORM is filled with values of *MP* class, and the documents are retrieved from the deep web stored in the repositories for processing.

```
If (Core field = CF₁)
{       If (Allied feild-2 = AF2₁) then

            MP = AF1₁, LP = AF1₂
       Else If (Allied feild-2 = AF2₂) then
            MP = AF1₂, LP = AF1₁, AF1₃
       Else If (Allied feild-2 = AF2₃) then

            MP = AF1₃, LP = AF1₂, AF1₄
       Else If (Allied feild-2 = AF2₄) then

            MP = AF1₄, LP = AF1₃, AF1₅
       Else If (Allied feild-2 = AF2₅) then

            MP = AF1₅, LP = AF1₄, AF1₆
       Else If (Allied feild-2 = AF2₆) then

            MP = AF1₆, LP = AF1₅

       ..........
       Else If (Allied feild-2 = AF2ₘ) then
            MP = AF1ₙ, LP = AF1ₙ₋₁, AF1ₙ₊₁
       Else

            (MPs and LPs does not exist at the
            time)        }
```

1.6 Algorithm

The main objective of combining the indexer within the proposed crawler is to uses the ideal time of the crawler. The HTML parser removes various TAGS present in the fetched HTML documents. As a result, the web page is segmented into set of words, where the stop words are removed as well as stemming of keywords is carried out to construct the inverted index. The inverted index stores the term along with its frequency of occurrence. Crawler components communicate with each other through web services (Chakrabarti et al. 1999).

```
1. Input URL and parse the HTML documents

2. Check for the FORM existence

3. If FORM does not exist then
      a. URLs are given to SURFACE WEB CRAWLER to fetch
         the document
      b. Downloaded document is processed by INDEXER to
         update inverted index list

4. Else
      a. URLs are given to DEEP WEB CRAWLER
      b. The FORM is analysed
      c. The FORM is processed by populating LVS table
      d. The Core field and Allied Fields from LVS ta-
         ble are identified depending on their func-
         tional dependencies based on the application
         domain.
      e. AF1_{i(y)MP}=(AF2_{j(x)}-1)+CF_i gives Most Preferable
         class, AF1_{i(y)LPT}=(AF2_{j(x)}+1)-1+CF_i gives adjacent
         top  Least  Preferable  class  and
         AF1_{i(y)LPD}=(AF2_{j(x)}-1)-1+CF_i gives adjacent down
         Least Preferable class.
      f. The above theoretical foundation with Subset
         of Rules are given to SVM classifier
      g. Rest of the Rules are generated by the classi-
         fier which classifies MPs and LPs for each
         Core field
      h. The FORM is Filled with MP values and the doc-
         uments are fetched
      i. Data is stored in repository and processed by
         INDEXER to update inverted index list
```

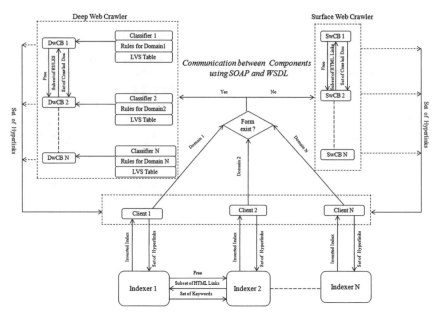

Fig. 1.4 Functional view of distributed web crawler

1.7   Functional Block Diagram of Distributed Web Crawlers

In this subsection, functional block of distributed crawler is explained. HTMP parser uses web service-based communication mechanism. This communication standard sends XML information between the distributed crawler. Figure 1.4 depicts the functional behaviour of the distributed crawler. The proposed crawler is scalable to accommodate more number of crawler, and due to space constraint, only two crawlers are shown in Fig. 1.4

Various components of the crawler communicate using XML messages. The load of each component is continuously monitored in terms of XML messages. While the load of a component reaches beyond a threshold, the other free component of the architecture is assigned the job.

1.8   Experimental Results

For experimental purpose, the real estate domain is fixed as web applications and considered http://www.99acres.com. The website has large number of FORMs, where each element corresponds to finite set of values. For example, input elements, such as combo box and list box, have to be filled by the users with predefined values.

Table 1.1 Estimated relationship between label and value for real estate web application

Property type	Budget	City
New projects	Below 5 Lacks	Delhi/NCR (All)
Residential apartment	5 Lacks	Delhi Central
Independent/builder floor	10 Lacks	Delhi Dwarka
Independent house/villa	15 Lacks	Delhi East
Residential land	20 Lacks	Delhi North
All residential	25 Lacks	Delhi Others
Studio apartment	30 Lacks	Delhi South
Farmhouse	40 Lacks	Delhi West
Serviced apartments	50 Lacks	Faridabad
Other	60 Lacks	Ghaziabad
All commercial	75 Lacks	Greater Noida
Commercial shops	90 Lacks	Gorgon
Commercial showrooms	1 Crore	Noida

Table 1.2 Combination of $AF2_j$ and $AF1$ for real estate web applications

Cities (CF_1)	Price set ($AF2_j$) for the apartment bedrooms ($AF1_i$)					
	1 BR	2 BR	3 BR	4 BR	5 BR	6 BR
A1	40 Lacks	75 Lacks	1 Crore	1.5 Crore	2 Crore	3 Crore
A	30 Lacks	50 Lacks	75 Lacks	1 Crore	2 Crore	2.5 Crore
B1	40 lacks	60 Lacks	1 Crore	1.5 Crore	2 Crore	2.5 Crore
B2	30 Lacks	40 Lacks	80 Lacks	1 Crore	No property exist at the time	No property exist at the time
C	30 Lacks	45 Lacks	60 Lacks	80 Lacks	No property exist at the time	No property exist at the time

These filled element-value combination is used for constructing LVS, which is shown in Table 1.1. It is observed from the table that most of the important fields are mapped as primary fields. In this case, the city is an important filed and is classified as A_1, A, B_1, B_2, C. The property type, number of bedrooms and price range are considered as allied ones. In the given example, number of bedrooms is categorized as six groups [B_1–B_6], and price is categorized as eight groups [R_1–R_8].

Table 1.2 shows the price ranges ($AF2_j$), by which most of the relevant web pages are retrieved for bedrooms ($AF1_i$).

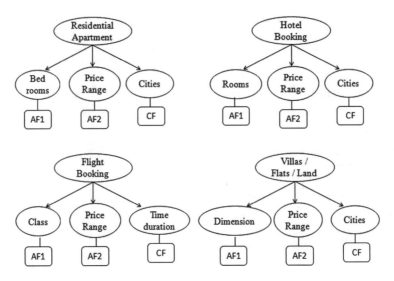

Fig. 1.5 Sample core and allied fields

The logical view of core and allied field is shown in Fig. 1.5 for a real estate and travel domain. The content of Table 1.2 is processed for generating the rules by applying suitable classification scheme. Similarly, Eqs. (1.2–1.4) generate rules for http://www.99acres.com which are presented below.

1.8.1 Rules for Real Estate Domain in http://www.99acres.com

The FORM input elements and value combination from MP class alone is chosen and found that a large number of relevant documents are fetched from database applications. In contrast, the rules of least preferable and mutually exclusive class may not fetch relevant documents from FORM-based web applications. The sample rule and the SVM-based classification result of most preferable and least preferable classes are depicted in Fig. 1.6. For different core fields, budget of the property and type of the property are considered as allied fields. The rules of most preferable and other classes are classified using the linear kernel function.

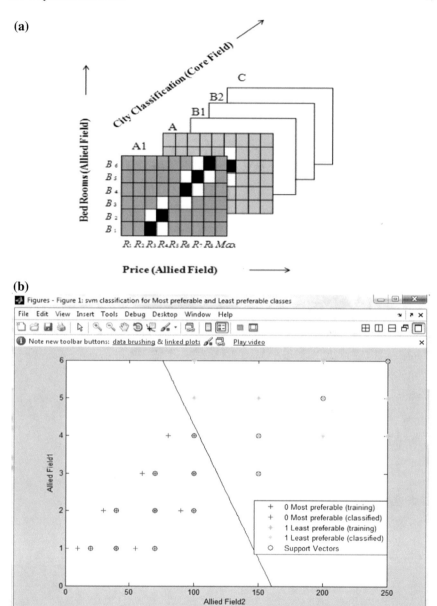

Fig. 1.6 **a** Rule for real estate web application. **b** Classification result

```
If (City = A1)
{      If (Price=R₃) then MP= B₁, LP= B₂
       Else If (Price=R₄) then MP= B₂, LP= B₁, B₃
       Else If (Price=R₆) then MP= B₄, LP= B₃, B₅
       Else If (Price=R₇) then MP= B₅, LP= B₄, B₆
       Else If (Price=R₈) then MP= B₆, LP= B₅
       Else (Property does not Exist)        }

If (City = A)
{      If (Price=R₂) then MP= B₁, LP= B₂
       Else If (Price=R₃) then MP= B₂, LP= B₁, B₃
       Else If (Price=R₄) then MP= B₃, LP= B₂, B₄
       Else If (Price=R₆) then MP= B₄, LP= B₃, B₅
       Else (Property does not Exist)        }

If (City = B1)
{      If (Price=R₃) then MP= B₃, B₄, LP= B₂, B₁, B₆
       Else If (Price=R₄) then MP= B₂, LP= B₁, B₃
       Else (Property does not Exist)        }

If (City = B2)
{      If (Price=R₁) then MP= B₁, LP= B₂
       Else If (Price=R₂) then MP= B₂, LP= B₁, B₃
       Else If (Price=R₃) then MP= B₃, LP= B₂, B₄
       Else If (Price=R₄) then MP= B₄, LP= B₃, B₅
       Else (Property does not Exist)        }

If (City = C)
{      If (Price=R₂) then MP= B₁ B₂ B₃ B₄, LP= B₅ B₆
       Else (Property does not Exist)        }

If (City = Others)
{      If (Price= Max) then MP= B₁ B₂ B₃ B₄ B₅ B₆
       Else (Property does not Exist)        }
```

1.8.2 Performance of Deep Web Crawler

In this work, the precision of retrieval is used as performance metric by which the amount of relevant web data fetched is calculated. The precision of retrieval for core and allied field with respect to the real estate domain is presented in Fig. 1.7. The values of the FORM belonging to MP classes are filled, and the documents are retrieved, and the inverted index is updated accordingly. For each city, we have

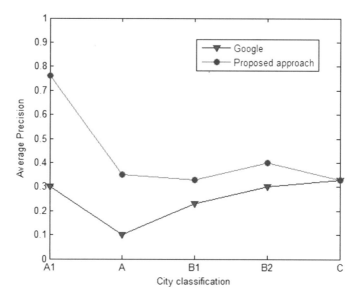

Fig. 1.7 Average precision

selected 100 queries and submitted as query in the proposed crawler and Google (http://www.Google.com). The precision of retrieval achieved by the crawler is more than that of Google system.

The coverage is yet another metric to measure the performance of the crawler. It is defined as the number of crawled document to the total number of document present in the web database as shown in Eq. (1.5). Equations (1.5) and (1.6) denote the coverage rate.

$$CR(q_i, DB) = \left(\frac{|C(q_i, DB)|}{|DB|} \right) \tag{1.5}$$

In the above equation, $|C(q_i, DB)|$ is the number of crawled document from web database for query q_i and $|DB|$ is the total number of records present in web database.

The coverage rate for a set of queries $Q = \{q_1, q_2, \ldots, q_n\}$ is defined as

$$CR(q_i, \vee \ldots \ldots \vee, q_n, DB) = \left(\frac{|C(q_1, DB) \cup \ldots \ldots \cup C(q_n, DB)|}{|DB|} \right) \tag{1.6}$$

where $|C(q_1, DB) \cup \ldots \cup C(q_n, DB)|$ the number of the union of the records in DB is matched q_i.

Figure 1.8 presents the coverage rate for http://www.99acres.com, http://www.in diaproperty.com and http://www.makemytrip.com. In Fig. 1.8a, the coverage rate is presented. The performance of both is similar. This is due to the fact that Google periodically updates its database, and in contrast, the proposed updates once.

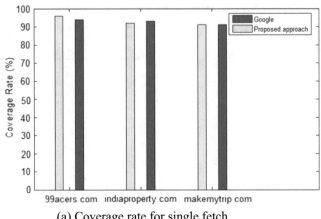

(a) Coverage rate for single fetch

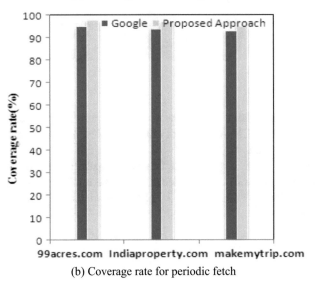

(b) Coverage rate for periodic fetch

Fig. 1.8 Coverage rate of the crawler. **a** Coverage rate for single fetch. **b** Coverage rate for periodic fetch

To handle this issue, the proposed crawler has periodically updated database and measured the coverage rate and is presented in Fig. 1.8b. It is noticed that the performance is above 90% compared to other search engine system.

The experiments are carried out on a computer having Intel(R) Xeon(R) CPU at 2.40 GHz with 12.0 GB RAM configuration, and the implementation language is C#. It is noticed that deep web crawler achieves higher performance metrics compared to the surface web crawler. The performance of deep and surface web crawler is compared and presented in Fig. 1.9. For a time duration, both the crawlers have fetched the documents from http://www.99acres.com, and the number of documents

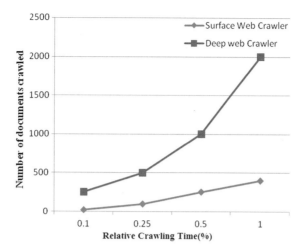

Fig. 1.9 Retrieval rate by crawlers

Table 1.3 Coverage Ratio

Keywords	Number of documents crawled produced in response		
	Surface web crawler	Deep web crawler	Ratio
Gorgon	2	546.23	1:274
Noida	4	748.15	1:188
Hyderabad	5	1946	1:389
Residential	21	2016	1:96
Apartment	15	2013	1:134

fetched by each crawler is measured. It is evident that deep web crawler fetched large number of documents compared to the surface web crawler. This is due to fact that the crawler has filled the FORMs with values of most preferable class.

In addition to the above experiment, the performance of the deep and surface web crawler is measured based on users query submitted in the query interface of the web applications. It is noticed from Table 1.3 that the performance of deep web crawler is good compared to surface web crawler. The deep crawler has retrieved 2000 more web pages, and this is again due to the values belonging to Most Preferable class.

1.9 Conclusion

The architecture specification of a surface, deep web crawler with indexer is presented. Rules are derived in the form of if-then-else construct for filling the HTML FORMs. The fields= of FORMs are classified as core and allied fields by which the FORMs are filled effectively. The FORMs element values are classified as Most

Preferable, Least Preferable and Mutually Exclusive. The FORMs are filled with values belonging to the Most Preferable classes to achieve the higher coverage rate. The proposed crawler is experimentally tested on real estate websites and found that it has fetched more number of relevant documents. The distributed specification of the crawler uses web services and XML message to use the ideal time of other crawler. The performance is compared with well-known retrieval systems and found that the proposed crawler is performing well. In future, the LVS table will be dynamically populated for various domains by automating rule generation process. The current system handles only the finite domain elements values of the FORM and will be scaled to handle the infinite domain values.

References

Ajoudanian, S., & Jazi, M. D. (2009). Deep web content mining. *Proceedings of World Academy of Science: Engineering and Technology, 49.*

Alvarez, M., Raposo, J., Pan, A., Cacheda, F., Bellas, F., & Carneiro, V. (2007). Crawling the content hidden behind web forms. In *Computational Science and Its Applications Proceedings of the International Conference* (Part II, pp 322–333). Berlin, Heidelberg: Springer.

Arasu, A., Cho, J., Garcia-Molina, H., & Raghavan, S. (2001). Searching the web. *ACM Transactions on Internet Technologies, 1*(1), 2–43.

Barbosa, L., & Freire, J. (2004). Siphoning hidden-web data through keyword-based interfaces. In *XIX Simpsio Brasileiro de Bancos de Dados* (pp. 309–321).

Barbosa, L., & Freire, J. (2007). An adaptive crawler for locating hidden web entry points. In *World Wide Web Proceedings of the 16th International Conference* (pp. 441–450). New York, NY, USA: ACM.

Brooks, T. A. (2004). The nature of meaning in the age of Google. *Proceedings of Information Research, 9*(3).

Caverlee, J., Liu, L., & Rocco, D. (2006). Discovering interesting relationships among deep web databases: A source-biased approach. *Journal of World Wide Web, 9*(4), 585–622.

Chakrabarti, S., Dom, B., Kumar, R., Raghavan, P., Rajagopalan, S., Tomkins, A., et al. (1999). Mining the Web's link structure. *Computer, 32*(8), 60–67.

Chang, K. C.-C., He, B., Li, C., Patel, M., & Zhang, Z. (2004). Structured databases on the web: Observations and implications. *SIGMOD Record, 33*(3), 61–70.

Gianvecchio, S., Xie, M., Wu, Z., & Wang, H. (2008). Measurement and classification of humans and bots in internet chat. In *Proceedings of the 17th International Conference on Security Symposium*, Association Berkeley, USA (pp. 155–169).

Kayed, M., & Chang, C.-H. (2010). FiVaTech: Page-level web data extraction from template pages. *IEEE Transactions on Knowledge and Data Engineering, 22*(2), 249–263.

Liu, J., Lu, J., Wu, Z., & Zheng, Q. (2011). Deep web adaptive crawling based on minimum executable pattern. *Journal of Intelligent Information Systems, 36*(2), 197–215.

Madhavan, J., Ko, D., Kot, L., Ganapathy, V., Rasmussen, A., & Halevy, A. (2008). Google's deep web crawl. *Proceedings of the VLDB Endowment, 1*(2), 1241–1252.

Ntoulas, A., Zerfos, P., & Cho, J. (2005). Downloading textual hidden web content through keyword queries. In *ACM/IEEE-CS Proceedings of the 5th Joint Conference on Digital Libraries* (pp. 100–109). New York, NY, USA: ACM.

Ntoulas, A., Zerfos, P., & Cho, J. (2008). *Downloading hidden web content.* UCLA, Computer Science. Retrieved February 24, 2009.

Raghavan, S., & Garcia-Molina, H. (2001). Crawling the hidden web. In *Very Large Databases (VLDB F01) Proceedings of the 27th International Conference* (pp. 129–138). San Francisco, CA, USA: Morgan Kaufmann Publishers Inc.

Rennie, J., & McCallum, A. (1999). Using reinforcement learning to spider the web efficiently. In *Machine Learning (ICML) Proceedings of the 16th International Conference* (pp. 335–343). San Francisco, CA, USA: Morgan Kaufmann Publishers Inc.

Wei, L., Xiaofeng, M., & Weiyi, M. (2010). ViDE: A vision-based approach for deep web data extraction. *IEEE Transactions on Knowledge and Data Engineering, 22*(3), 447–460.

Wu, P., Wen, J. R., Liu, H., & Ma, W. Y. (2006). Query selection techniques for efficient crawling of structured web sources. In *Data Engineering Proceedings of the 22nd International Conference*, Atlanta, 2006 (pp. 47–56).

Yongquan, D., & Qingzhong, L. (2012). A deep web crawling approach based on query harvest model. *Journal of Computational Information Systems, 8*(3), 973–981.

Zhao, P., Huang, L., Fang, W., & Cui, Z. (2008). Organizing structured deep web by clustering query interfaces link graph. In *Advanced Data Mining and Applications Proceedings of the 4th International Conference of ADMA '08* (pp. 683–690). Berlin, Heidelberg: Springer.

Chapter 2
Information Classification and Organization Using Neuro-Fuzzy Model for Event Pattern Retrieval

2.1 Introduction

Cognitive scientists have believed that people experience and cognize the objective world based on various Events (Zhou et al. 2008). The information related to these Events is huge and updated over the Internet in various forms. It is observed that the Events in Web documents occur every time with different degrees of significance related to some place/person. The Event describes global happenings that are occurring at a specific time and place. This encourages researchers to sense the real world by extracting knowledge on topics, Events, stories from a large volume of Web data (Allan et al. 2003). These topics or Events can be further used in various applications, say automatic Question Answering (QA), Text Summarization (TS) and Semantic Web Retrieval (SWR). However, it is found that the information associated with various Events is unstructured and it is difficult to classify and extract the useful and interested Event information. Hence, a system with a suitable approach is required that quickly understands the documents, extracts the information and exhibits their structure by highlighting the possible subsets of interest. The unstructured text, which includes facts and Events, needs to be preprocessed for extracting usable structured data. Even after extracting, the obtained text is raw and the patterns of the sentence/text need to be analysed for identifying interested and useful Event information. It is well-known that analysing and classifying the sentence patterns for Events is a challenging task. Even well-known Web search approaches may not fully satisfy the user's information requirement even though these approaches are fast in locating and extracting the information. This is due to the lack of information in understanding the sentence patterns. As a result, understanding, identifying and classifying the sentence patterns are significant for extracting interested and useful information about Events for an application domain.

In natural language, the semantic portion of the sentences is composed of nouns, verbs, prepositions and other Parts Of Speech (POS) tags, and all of them constitute some Event Type. In addition, semantic contents, temporal proximities and named

© Springer Nature Singapore Pte Ltd. 2018
S. G. Shaila and A. Vadivel, *Textual and Visual Information Retrieval using Query Refinement and Pattern Analysis*, https://doi.org/10.1007/978-981-13-2559-5_2

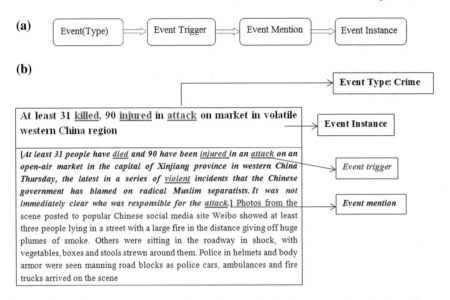

Fig. 2.1 Event features. **a** Relationship of Event features, **b** Sample Event features for crime-related document

entity recognitions are also considered for detecting Events. For an Event (Type), the sentences in Web document contain a large number of terms and only certain terms represent the Event (depends on applications). The terms that represent the Event are called Event Trigger terms, and sentences that describe the Event are termed as Event Mentions. As a result, an Event can have multiple Mentions associated with it. Similarly, a specific occurrence of a particular Event is referred to as Event Instance. The Event Mention in the text clearly expresses the occurrence of the Event Instance. For example, '*nearly 3000 people were killed by terror attack on WTC on 9/11*' is considered as Event Instance. An Event Instance is something that happens at a specific time and place. All these Event features are related to each other and contribute to Event Type as shown in Fig. 2.1a.

The relation between various Event features in the text document is shown in Fig. 2.1b for crime domain. In the crime domain, the Web documents have information related to crime Events. The text information in the document consists of Event Triggered terms such as *kill* and *attack*. The Event Mention sentences associated with these Event Triggered terms express an Event Instance in the form of Event happenings that has occurred at specific time and place.

Based on the above discussion, the presented research is motivated as follows:

- To understand the importance of term features in the sentences
- To understand the POS nature of the sentences for classification
- To analyse the classified patterns for useful and interested information about Event Instances
- Finally, to build the Event Corpus with the Instance patterns.

In this chapter, Event detection is considered as a pattern identification and classification problem at the sentence level. Based on the above-mentioned objectives, the sentence patterns are identified for interested and useful information about Event Instances. For a given Event Type, the sentence nature, Event Triggered terms along with its immediate co-occurrence terms are considered as features for classifying the sentences. The classification rules are generated based on the POS tags of these features, which yield eight Event Mention sentence patterns. It is observed that some of these patterns have more interested and useful information about Event Instances. Due to the constraint of only two term features (Event Triggered and immediate co-occurrence terms) used for classification, important sentences are misclassified. It is found that the boundary of the Event Mention sentence patterns overlaps with each other and is handled by fuzzy tool. A set of fuzzy rules are generated by giving importance for all term features in the sentence for obtaining more number of Event Mention patterns. These patterns are assigned weights based on the type and quality of information content for understanding the real-world Event Instances. Finally, Event Type Corpus is built by extracting only high-weighted patterns, which is further used for any applications. The approach is domain specific and can be used for retrieving the domain-specific interested information. The rest of the chapter is organized as follows. The chapter begins with a brief description of background research in Sect. 2.1. Sections 2.2–2.7 describe the presented approach for classifying the sentence patterns that give interested and useful information about Instances for an Event Type. The performance of Event detection approaches in Sect. 2.8 is experimentally evaluated and concluded the chapter in the last section of the chapter.

2.2 Reviews on Event Detection Techniques at Sentence Level

Detecting, classifying and extracting the Events from WWW have engaged the attention of the researchers, and many new approaches have been reported during the last few decades. In 1998, the National Institute of Standards and Technology (NIST) sponsored a project named Topic Detection and Tracking (TDT), which has focused on development of technologies that could recognize new Events in news stream and track the progression of these Events over time (Yang et al. 1998; Allan et al. 1998). This project was wound up during 2004, and later, it was continued by Automatic Content Extraction (ACE) program (http://www.ldc.upenn.edu). Both the Message Understanding Conference (MUC) (Grishman and Sundheim 1996) and Automatic Content Extraction (ACE) program have been initiated to focus primarily on Event detection. The MUC has been developed for automatically identifying and analysing military messages present in textual information. The main objective of the ACE program is to categorize the popular Events from news articles into various classes and sub-classes. Previous ACE and TDT researches have focused on Event detection at the term/phrase level and the document level; however, no prior work has been done at sentence level for Event detection. Apart from these projects, many approaches such as Integrated Web Event Detector (IWED) (Zhao et al. 2006a, b) have been proposed to extract Events from Web data by integrating author-centric data and visitor-centric data. The Web data is modelled as a multi-graph, each vertex is mapped with a Web page, and each edge represents the relationship between connected Web pages in terms of structure, semantic and usage pattern. The Events are detected by extracting strongly connected sub-graphs from the multi-graph using a normalized graph cut algorithm. However, the approach was able to detect Events to certain extent only due to the problem in identifying exactly the strongly connected sub-graphs. An approach has been proposed for detecting Events using click-through data and the log data of Web search engines (Zhao et al. 2006a, b). The click-through data is first segmented into a sequence of bipartite graphs based on user-defined time granularity. The sequence of bipartite graphs is represented as a vector-based graph and is used for maintaining the semantic relationships between queries and pages. However, the approach could not extract the Event Instance semantic patterns efficiently due to the subjective nature of time granularity. The approach in Hung et al. (2010) extracts interesting patterns by matching the text with Web search query results. The lexico-syntactic patterns and semantic role labelling technique is applied to parse the extracted sentences and essential knowledge associated with the Events that are identified in each sentence. In Cohen et al. (2009), the authors have proposed an approach to recognize the Event in the biomedical domain as one of the means for conceptual recognition and analysis. A method has identified the concepts using named entity recognition task. Automatically identifying Event extent, Event Trigger and Event argument using the bootstrapping method (Xu et al. 2006) have been proposed. Nobel Prize-winning domain has been used to identify the Events by extracting rules from the text fragments using binary relations as seeds. However, the binary relation is crisp and vague and has ambiguity in the text fragments, and thus, the semantic information is not fully captured.

Scalable Event detection (Aone and Ramos-Santacruz 2000) has been proposed that relies on text matching between sentences and a bag of predefined lexico-syntactic patterns. The syntactic and semantic restrictions are specified in the verb argument for the target Event. The lexico-syntactic patterns are obtained manually from examples collected by knowledge experts. However, this approach generates rules manually using a large number of patterns, and it is difficult to control these rules for different Events since there are too many patterns among various Events. In addition, it is tedious to have complete list of syntactic patterns for achieving encouraging recall rate. A two-phase novel topic detection algorithm (Yang et al. 2002) has been developed, where the incoming news has been classified into predefined categories. The topic-conditioned heuristics have been used to identify the new Events. Though the heuristics are found to be useful, the time information is not considered, resulting poor performance in Event detection. Retrospective new Event Detection (RED) (Hristidis et al. 2006) has been proposed for handling above issue, and it identifies Events in the news Corpus by taking both the content and the time information. However, the approach failed to exactly identify the semantic patterns that contribute for Event Instances. With the popularity of Web search engines, a huge amount of click-through data has been accumulated and available for analysis. The statistical and knowledge-based technique (McCracken et al. 2006) have been used for extracting Events by focusing on the summary report of a person's life incidents to extract the factual accounting of incidents. However, one of the drawbacks of this approach is that it covers entire Instance of each and every verb in the Corpus. As a result, the approach may not identify the Instance types associated with other POS. Various approaches have been proposed by considering every Event in a small portion. An approach has been proposed in Abuleil (2007), where the Events are detected and extracted by breaking each Event as elements by understanding the syntax of each element. Suitable classifier has been used for classifying each task into two classes, namely memory-based learners and maximum-entropy learners. It is noticed that by considering an Event at microlevel also may not help to exactly identify Events. A language modelling technique has been proposed in Naughton et al. (2010), where the Events are modelled as an On-Event and Off-Event and are classified using SVM classifier. However, the classification accuracy is low for rich features yielding 90% as Off-Event and 10% as On-Event sentences. This issue has been handled in Makoto et al. (2010), where an automatic Event extraction system has been presented by solving a classification problem with rich features.

It is noticed from the above discussion that most of the approaches detect Events without understanding the patterns of the text and have been handled in David (2006). In this approach, the Event detection approach is divided into four classification problems, namely trigger identification, argument identification, attribute assignment and Event co-reference resolution. The lexical, WordNet and dependency tree features are used, and argument identification is treated as a pair-wise classification problem. A separate classifier has been used for training each attribute, and finally, Event co-reference is treated as a binary classification task. In Grishman and Sundheim (1996), user specifies an Event of interest using several keywords as a query. The response to the query is a combination of streams, say news feeds and emails that are sufficiently correlated. They collectively contain all query keywords within a time period. A supervised method (Creswell et al. 2006) has been developed for detecting nominal Event Mention and formulated techniques to detect specific Events in unstructured text. The method has been used in Event-based common sense extraction (Filatova & Hatzivassiloglou 2004). However, it is laborious to apply them in more generic heterogeneous platform. This is due to the fact that there is a lack of labelled sentences or documents for Events with a widely varying scope.

Based on the above discussion, it is observed that for an Event Type, detecting the interested and useful information patterns about Instance at sentence level is most important and challenging. Since these Instance patterns have important information in the form of Events, this information can be retrieved and used for analysis, Event predictions, etc. Most of the above-mentioned approaches have failed in understanding the sentential patterns for capturing the knowledge content with respect to Event Instance. In this work, all the above issues are considered by classifying the sentence using a set of fuzzy rules and Event Corpus is constructed for retrieving useful information using Instance patterns.

2.3   Schematic View of Presented Event Detection Through Pattern Analysis

In this work, Event analysis is considered as pattern identification and classification problem and considered crime Event detection as an application domain. The flow of the concept is logically depicted in Fig. 2.2. It is noticed from Fig. 2.2 that the documents related to crime domain are crawled from WWW and inverted index is constructed. From the inverted index, only crime-related terms (along with synonyms, hyponyms, hypernyms) are extracted and maintained along with their sentence (s) in the sentence repository. The crime-related terms are referred as Event Trigger (et) terms, and the associated sentences are referred as Event Mention sentence. The Event Mention sentences are POS tagged for analysing the patterns to identify the useful and interested information about Event Instances. Based on the sentential nature (s), Event Mention sentences are initially classified as *Subjective* and *Objective* classes and further hierarchically classified based on POS tags of term features (et, $ct_{(I)}$). Here, $ct_{(I)}$ is the immediate co-occurrence term of et. The classification rules are validated using CART tool. It is observed that the rules are failed to clearly define the boundary between the patterns. As a result, ambiguity and impreciseness between the patterns exist. This affects the classification accuracy by leaving out large number of Event Instance patterns as outliers. This is due to the constraints of using only two term features (et, $ct_{(I)}$) of rules. Thus, as shown in Fig. 2.2 the rules are refined by considering all term features (s, et, $ct_{(I)}$, $ct_{(NI)}$). Here, $ct_{(NI)}$ is the non-immediate co-occurrence term of et. The rules are learnt by ANFIS model, which classifies sentences into sixteen Event Mention patterns. From the knowledge of previous hierarchical classification and based on the richness in the information content, linguistic grades are estimated for the newly obtained patterns. A suitable weighting mechanism is presented in such a way that the patterns with interested information are assigned higher weight and vice versa. An Event Type Corpus is built with higher weighted patterns and referred as Event Instance patterns. In subsequent sections, the presented approach is explained in detail.

2.4   Building Event Mention Sentence Repository from Inverted Index

A large number of crime-related Web documents are fetched from WWW with crime-related terms. In general, the terms in sentences are the basic unit of information and have semantic relationships within them. The terms are extracted from the documents, preprocessed by removing the stop words and stored in the inverted index. Stemming is avoided to retain linguistic patterns of the terms, which are necessary for POS tagging. For each term $<t>$ in the inverted index, there is a posting list that contains a document *id* and frequency of occurrence ($<d, f>$). Let D be a set of documents

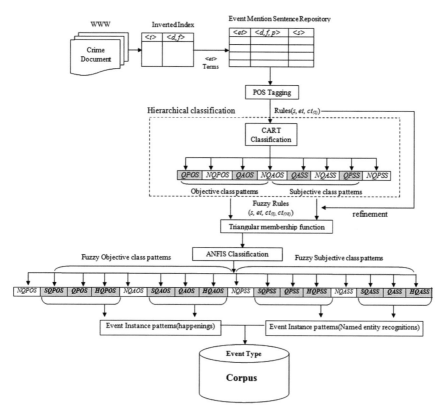

Fig. 2.2 Presented approach framework

crawled from WWW and T be a set of terms present in D. This may be treated as a labelling approach and denoted as follows.

$$l : T \times D \rightarrow \{True, \ False\} \tag{2.1}$$

The inverted index consists of crime-related terms as well as other terms. In this work, the crime-related terms are referred as Event Triggered (et) terms and set of the seed et terms are extracted from the inverted index. The synonyms, hyponyms and hypernyms of et terms are used which help in extracting rest of the et terms present in the inverted index. For instance, shoot, kill, death and murder are considered as et terms in crime-related Event. From Eq. (2.1), it is assumed that a term $t \in T$ present in a document $d \in D$, if $l : (t, d) = True$. In document retrieval applications, the posting list $<d, f>$ for a term $<t>$ is extracted from the inverted index, where f is term frequency in document d. Since the term $<t>$ is physically appearing in documents, given a set of et terms, such that $et \subseteq T$ can be written as a relationship of $<et, d, f>$ and is represented in Eq. (2.2).

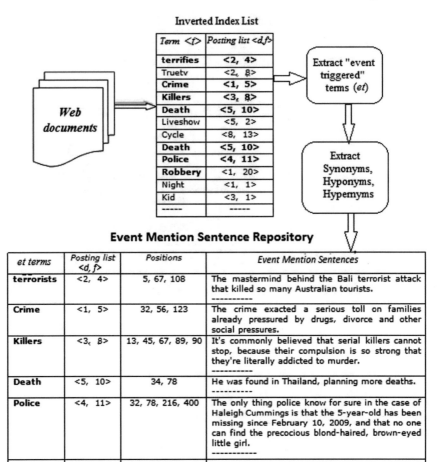

Fig. 2.3 Event Mention sentence repository

$$C^D(et) = \{< et, d, f > \,|d \in D, et \subseteq T \text{ and } l : (et, d) = True\} \qquad (2.2)$$

In Eq. (2.2), $C^D(et)$ is a posting list for et terms obtained in the form of $<et, d, f>$ with constraints $d \in D$, $et \subseteq T$ and $l : (et, d) = True$. The second part of Eq. (2.2) says that et terms are subset of terms T in inverted index and T belongs to documents D. Since $d \in D$, et should appear in d. Thus, all the sentences related to et terms are updated into repository and referred as Event Mention repository as depicted in Fig. 2.3. It is observed that the sentences in repository are huge with many Instances and describes crime Event using various sentence structures.

Further, the Event Mention sentences from the repository are fetched for POS tagging. The Stanford NLP tool is used as it is a maximum-entropy POS tagger for English and other languages. The tool reads text and assigns POS tags such as

Table 2.1 POS-tagged Event Mention sentences

et terms	Event Mention sentence POS tagging	POS Pattern
terrorists	The/DT mastermind/NN behind/IN the/DT Bali/NNP **terrorist/NN** attack/NN that/WDT killed/VBD so/RB many/JJ Australian/NNP tourists/NNS./.	DT/NN/IN/DT/NNP/**NN**/NN/WDT/VBD/RB /JJ /NNP/NNS./.
Crime	The/DT **crime/NN** exacted/VBD a/DT serious/JJ toll/NN on/IN families/NNS already/RB pressured/VBN by/IN drugs/ NNS./. divorce/NN and/CC other/JJ social/JJ pressures/NNS./.	DT/**NN**/VBD/DT/JJ/NN/IN/NNS/RB/VBN/ IN /NNS./. /NN/CC/JJ /JJ/NNS./.
Killers	It/PRP's /VBZ commonly/RB believed/VBN that/IN serial/JJ **killers/NNS** cannot/IN stop/NN ./, because/IN their/PRP compulsion/NN is/VBZ so/RB strong/JJ that/IN they/PRP 're/VBP literally/RB addicted/JJ to/TO murder/VB./.	PRP/VBZ/RB/VBN/IN/JJ/**NNS**/IN/NN,/,/I N/PRP/NN/VBZ/RB/JJ/IN/PRP/VBP/RB/JJ /TO/VB./.

noun, verb and adjective to each token word in the sentences. All the Event Mention sentences in the repository are POS tagged, and POS patterns are obtained. A sample Event Mentions and their POS patterns are represented in Table 2.1. For instance, '*terrorist*' is considered as an *et* term and its corresponding Event Mention sentence is POS tagged where the POS pattern for this term is a noun (*NN*). Finally, the obtained POS patterns of the repository are considered further for classification with detailed explanation in Sect. 2.5.

2.5 Event Mention Sentence Classification

As the size of the Event Mention sentence repository is huge and consists of different sentence structures, it is difficult to understand their POS patterns to extract the knowledge from them. However, POS tags play significant role in identifying the useful and interested information about Instances in the sentences. Hence, four POS tags such as verb, adverb, noun and adjective are considered along with their extensions for Event Mention sentence classification and are represented in Table 2.2.

Using POS tag features, the Event Mention sentences are classified hierarchically as shown in Fig. 2.4.

The first level of classification is based on the sentence nature, which classifies the sentences into *Subjective* class and *Objective* class. It is observed that some sentences have action verbs, which give detailed description about the activities/happenings

Table 2.2 POS tags used for sentence classification

POS tag	Extension POS tags
JJ	{JJ, JJR, JJS}
VB	{VB, VBZ, VBD, VBG, VBN, VBP}
NN	{NN, NNP, NNPS, NNS}
RB	{RB, RBR, RBS}

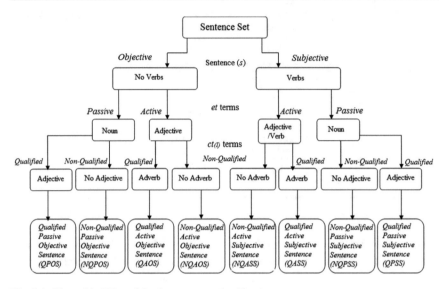

Fig. 2.4 Hierarchical Event Mention sentence classification

associated with the Event, and these sentences are referred as *Subjective* class. The *Subjective* class sentences in general are dynamic in nature and usually present in the *<body>* section of a Web page. This class triggers the Event through Event happenings and provides useful information. On the other hand, some sentences do not have action verbs and referred as *Objective* class. These sentences are static in nature and usually occur in *<title>*, *<anchor>* and *<h₁ … h₆>* section of Web pages. These sentences are also important as they highlight Events through named entity recognitions and are important in identifying Event Instances. At the second level, the Event Mention sentences are classified based on the POS nature of the *et* terms. The *et* terms are considered as information units and play a vital role in the context of sentential semantics with the Events. Usually, *et* terms in the sentences appear as noun/verb/adjective POS tags, based on which the sentences are classified as *Passive* class or *Active* class. In general, noun represents named entity recognition in terms of place/thing/person. If the *et* term in the sentence appears as noun, such sentences are referred as *Passive* class. On the other hand, if the *et* term in the sentences appears as verb/adjective, such sentences are referred as *Active* class. The *Active* class sentence expresses Event happenings by giving descriptive information. In the last level of classification, the immediate co-occurrence ($ct_{(I)}$) terms that are associated with *et* terms are considered. The $ct_{(I)}$ terms estimate the degree of useful and interested information about the Instance in the Event Mention sentence. The $ct_{(I)}$ terms in the sentence may or may not appear as adjective/adverb POS tags. Based on the presence/absence of these tags, the sentences are classified as *Qualified* class and *Non-Qualified* class. If the $ct_{(I)}$ term in the sentence appears as adjective/adverb, such sentences are referred as *Qualified* class, since the degree of useful and interested information pieces about the Instance is high in the Event Mention sentence. In the absence of adjective/adverb, the sentences are referred as *Non-Qualified* class.

The hierarchical classification scheme can be understood in a better way through an example. For instance, the sentence '*Al-Qaeda attack*' has the POS pattern '*Al-Qaeda/JJ attack/NN*'. At the first level, the sentence is classified into *Objective sentence*, as the sentence nature is non-verb. In the second level, the sentence is further classified into *Passive Objective sentence* as *et* term '*attack*' is noun and plays vital role in identifying Event crime. In the last level, the sentence is classified into *Qualified Passive Objective sentence* as $ct_{(I)}$ is adjective and gives useful and interested information about crime Instance '*Al-Qaeda attack*'. The classification strategy used the knowledge of the terms and their relationship with the sentence. The three-level sentential classification (s, et, $ct_{(I)}$) scheme provides eight patterns. The sentence classification constraint is presented in the form of rules, which are derived using POS tag patterns:

```
Rule 1: if (s=VB) && (et=NN) && (ct(I)=JJ) then C=QPSS
             else if (s=VB) && (et=NN) && (ct(I)≠JJ) then C=NQPSS

Rule 2: if (s=VB) && (et=JJ/VB) && (ct(I)=RB) then C=QASS
             else if (s=VB) && (et=JJ/VB) && (ct(I)≠RB) then C=NQASS

Rule 3: if (s≠VB) && (et=NN) && (ct(I)=JJ) then C=QPOS
             else if (s≠VB) && (et=NN) && (ct(I)≠JJ) then C=NQPOS

Rule 4: if (s≠VB) && (et=JJ) && (ct(I)=RB) then C=QAOS
             else if (s≠VB) && (et=JJ) && (ct(I)≠RB) then C=NQAOS
             else O=Outlier
```

where s—sentence, et—Event Triggered term, $ct_{(I)}$—immediate co-occurrence term, *VB*—verb, *NN*—noun, *JJ*—adjective, *RB*—adverb and *C*—class.

To validate the presented hierarchical classification model, the CART tool (Breiman et al. 1985) is used, which creates a binary decision tree that classifies the data into one of $2n$ linear regression models. The CART tool has advantages such as simple classification form, coping with any data structure, using conditional information usefully, invariant under transformations of variables, robust with respect to outliers, and estimates the misclassification rate. For experiment, crime Event dataset having 500 Web documents is considered. After removal of duplicates, 500 *et* terms are extracted from the inverted index, and 46,938 related Event Mention sentences are obtained. These sentences are classified into *Subjective* and *Objective* classes in the first level.

Table 2.3 Event Mention sentence classification in three levels using CART

Levels	Split	Left	Right
1	'Objective', 'Subjective'	32,263	14,675
2	'Active', 'Passive'	13,420	33,518
3	'Qualified', 'Non-Qualified'	34,866	12,072

Table 2.4 Event Mention patterns

Event Mention patterns	Sentence-Count
Qualified Passive Objective Sentence (QPOS)	8155 (17.37%)
Non-Qualified Passive Objective Sentence (NQPOS)	17,846 (38.02%)
Qualified Active Objective Sentence (QAOS)	997 (2.12%)
Non-Qualified Active Objective Sentence (NQAOS)	5265 (11.22%)
Qualified Passive Subjective Sentence (QPSS)	2506 (5.34%)
Non-Qualified Passive Subjective Sentence (NQPSS)	5011 (10.68%)
Qualified Active Subjective Sentence (QASS)	414 (0.88%)
Non-Qualified Active Subjective Sentence (NQASS)	6744 (14.37%)

The *Objective* class contains 32,263 sentences, and the *Subjective* class contains 14,675 sentences. Similarly, second level is classified into *Active* and *Passive* classes, and the last level is classified into *Qualified* and *Non-Qualified* classes. Table 2.3 presents the classification output of the CART with number of sentences in each class.

The output of the CART contains eight Event Mention patterns and is shown in Table 2.4. Among 46,938 Event Mention sentences, 2506 (5.34%) sentences are classified as *QPSS* pattern. Similarly, 5011 (10.68%) sentences are *NQPSS* pattern and so on. It is observed from Table 2.4 that *Qualified* sentence patterns are less in number. However, these patterns are important since they have useful information about the Event Instances. The hierarchical classification has used only two term features $(et, ct_{(I)})$ after first-level classification. As a result, the rules are unable to capture complete semantic relation between terms in the sentence, which resulted in misclassification. It is also noticed that some of the common features are shared between the patterns, which leads to poor identification of *Qualified* sentence patterns.

It is also found that the CART-based analysis cannot be further extended since it is unstable when combinations of variables are used. This limitation can be handled by fuzzy logic approach. The problem due to lack of clear boundary between the patterns can be handled by fuzzy rules. Here, all the term features, i.e. s, et, $ct_{(I)}$ and

Table 2.5 Sentential term features and POS tags

Features	Sentence (s)
	Event Triggered term (et)
	Immediate Co-occurrence term ($ct_{(I)}$)
	Non-Immediate Co-occurrence term ($ct_{(NI)}$)
POS tags	Verb (VB)
	Noun (NN)
	Adjective (JJ)
	Adverb (RB)

$ct_{(NI)}$ (non-immediate co-occurrence) terms of the sentence, are considered. Thus, all the features are considered and presented in Table 2.5 for refining the rules to handle all the above-mentioned issues.

A set of fuzzy rules are derived, and artificial neuro-fuzzy inference system (ANFIS) (Jang 1993) is used to create a fuzzy decision tree to classify the patterns into one of $2n$ linear regression model. ANFIS considers the best features of fuzzy systems and neural networks, which integrates the advantages of smoothness due to the fuzzy interpolation and adaptability due to the neural network back propagation. The model captures the fuzzy nature of classes through its learning rule and adaptive nature. While constructing fuzzy rules, importance is given to all the term features (et, $ct_{(I)}$, $ct_{(NI)}$) in the sentence (s). The following section presents the fuzzy rules constructed using all the features of Table 2.5.

2.6 Refining and Extending the Rules Using Fuzzy Approach

In this work, the fuzziness in the pattern boundary is captured by a suitable membership function, and fuzzy if-then rules are constructed for refining and extending the hierarchical classification rules presented in Sect. 2.5. The $C = n$ pattern classification method for fuzzy if-then rule is used and is shown in Eq. (2.3).

$$\text{Rule } R_j : \text{If } x_1 \text{ is } S_{j1}, x_2 \text{ is } S_{j2}, \ldots \text{ and } x_n \text{ is } S_{jn},$$
$$\text{then class } C_j \text{ with } CF_j; \quad j = 1, 2, \ldots, N \tag{2.3}$$

In the above Eq. (2.3), R_j is the label of the jth fuzzy if-then rule, x_1, x_2, ..., x_n are sentential term features, S_{j1}, S_{j2}, ... , S_{jn} are POS variables, which represent antecedent of fuzzy sets, C_j is the consequent pattern, and CF_j is the certainty or linguistic grade of the fuzzy if-then rule R_j. The hierarchical rules presented in Sect. 2.5 are refined using fuzzy rules by deriving $ct_{(NI)}$ terms. For example, a **Rule** 1 in previous section for *QPSS* is limited to $s = VB$ && $et = NN$ && $ct_{(I)} = JJ$. For this rule, POS tag (adjective/adverb) is searched in $ct_{(NI)}$ terms and is extended. The presence/absence (true/false values) of respective POS tag for $ct_{(NI)}$ generates two rules. The absence (false) of respective POS tag for $ct_{(NI)}$ generates new refined rule ($s = VB$ && $et = NN$ && $ct_{(I)} = JJ$ && $ct_{(NI)} \neq JJ$) for *QPSS* pattern. Similarly, the presence (true) of respective POS tag for $ct_{(NI)}$ generates new pattern *HQPSS* with the rule ($s = VB$ && $et = NN$ && $ct_{(I)} = JJ$ && $ct_{(NI)} = JJ$). Likewise, all the hierarchical rules that belongs to *Subjective* and *Objective* classes can be refined by introducing $ct_{(NI)}$. Based on the presence/absence (true/false values) of POS value for $ct_{(NI)}$, the rules are differentiated and new patterns are obtained. The linguistic or certainty grades are estimated for the patterns using the previous knowledge of hierarchical classification. These grades represent the type and amount of useful information present in the patterns with respect to Event Instance. Both *Subjective* and *Objective* classes play individual role in representing the type of information, and *Poor*, *Medium*, *High* represent the degree of useful information in the form of Linguistic grades. The fuzzy rules along with new patterns and the estimated linguistic grades of both *Subjective* and *Objective* classes are represented in Table 2.6).

In general, rules are constructed with a theoretical base or based on expertise in application domain. The above shown fuzzy rules are derived based on the structure of sentences and POS tags. These rules can be verified using rule evaluation parameters such as incompleteness, inconsistency, circularity and redundancy. The detailed verification process is given in Appendix. It is found that the presented rules are not having anomalies. In addition, the usefulness of the rules derived in this work is evaluated in the experiment section and found that the Event is identified effectively at the sentence level and achieves higher accuracy, precision, recall and F1 score.

2.6.1 Membership Function for Fuzzy Rules

To further consolidate the fuzzy rules, a suitable well-known triangular fuzzy membership function is used. The reason is that the well-known fuzzy membership functions are proven and if the rules presented can be fit in with membership function, it is imperative that the presented rules are perfect. For spacing and clarity, the rules of *Subjective* class alone are considered and explained. It is known that *Subjective* class consists of *Active* and *Passive* sub-classes, among them *Active* sub-class plays a significant role in providing useful and interested information. The significance is mapped by assigning membership values for the patterns in the range of [0–1]. The membership values for the *Subjective Active* class patterns are in the higher range of [0.25–1] and for *Subjective Passive* class patterns, the membership values are in the lower range of [0–0.24]. This is demonstrated by considering a sample pattern *HQASS*. The rule for *HQASS* is defined using antecedent fuzzy set variables a, b, c as shown in Eq. (2.4). The membership function evaluates the fuzzy set variables of Eq. (2.4) by giving all test cases. Each test case obtains different patterns, and the membership values are assigned to patterns. This is represented in Eq. (2.5).

Table 2.6   Fuzzy rules of patterns for (a) *Subjective* class, (b) *Objective* class

Subjective class patterns	Fuzzy rules	Linguistic (CF) grades
NQPSS	$s = VB$ && $et = NN$ && $ct_{(I)} \neq JJ$ && $ct_{(NI)} \neq JJ$	Subjective Very Poor (SVP)
Semi-QPSS (SQPSS)	$s = VB$ && $et = NN$ && $ct_{(I)} \neq JJ$ && $ct_{(NI)} = JJ$	Subjective Poor (SP)
QPSS	$s = VB$ && $et = NN$ && $ct_{(I)} = JJ$ && $ct_{(NI)} \neq JJ$	Subjective Very Low (SVL)
High-QPSS (HQPSS)	$s = VB$ && $et = NN$ && $ct_{(I)} = JJ$ && $ct_{(NI)} = JJ$	Subjective Low (SL)
NQASS	$s = VB$ && $et = JJ/VB$ && $ct_{(I)} \neq RB$ && $ct_{(NI)} \neq RB$	Subjective Lower Medium (SLM)
Semi-QASS (SQASS)	$s = VB$ && $et = JJ/VB$ && $ct_{(I)} \neq RB$ && $ct_{(NI)} = RB$	Subjective Medium (SM)
QASS	$s = VB$ && $et = JJ/VB$ && $ct_{(I)} = RB$ && $ct_{(NI)} \neq RB$	Subjective Higher Medium(SHM)
High-QASS (HQASS)	$s = VB$ && $et = JJ/VB$ && $ct_{(I)} = RB$ && $ct_{(NI)} = RB$	Subjective High(SH)
Objective class patterns	Fuzzy rules	Linguistic (CF) grades
NQPOS	$s \neq VB$ && $et = NN$ && $ct_{(I)} \neq JJ$ && $ct_{(NI)} \neq JJ$	Objective Very Poor (OVP)
Semi-QPOS (SQPOS)	$s \neq VB$ && $et = NN$ && $ct_{(I)} \neq JJ$ && $ct_{(NI)} = JJ$	Objective Poor (OP)
QPOS	$s \neq VB$ && $et = NN$ && $ct_{(I)} = JJ$ && $ct_{(NI)} \neq JJ$	Objective Very Low (OVL)
High-QPOS (HQPOS)	$s \neq VB$ && $et = NN$ && $ct_{(I)} = JJ$ && $ct_{(NI)} = JJ$	Objective Low (OL)
NQAOS	$s \neq VB$ && $et = JJ$ && $ct_{(I)} \neq RB$ && $ct_{(NI)} \neq RB$	Objective Lower Medium (OLM)
Semi-QAOS (SQAOS)	$s \neq VB$ && $et = JJ$ && $ct_{(I)} \neq RB$ && $ct_{(NI)} = RB$	Objective Medium (OM)
QAOS	$s \neq VB$ && $et = JJ$ && $ct_{(I)} = RB$ && $ct_{(NI)} \neq RB$	Objective Higher Medium (OHM)
High-QAOS (HQAOS)	$s \neq VB$ && $et = JJ$ && $ct_{(I)} = RB$ && $ct_{(NI)} = RB$	Objective High (OH)

Fig. 2.5 Triangular membership function for *Subjective Active* class patterns

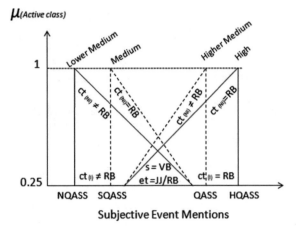

$$a = \begin{Bmatrix} s = VB \\ et = JJ/VB \end{Bmatrix}, \quad b = \{c_{(I)} = RB\}, \quad c = \{c_{(NI)} = RB\} \tag{2.4}$$

Evaluation of *mf (Subjective class)*:

1. *If (a=true), (b=true) and (c=true), then mf (HQASS)= 1*
2. *If (a=true), (b=true) and (c=false), then mf (QASS) ≥ 0.75*
3. *If (a=true), (b=false) and (c=true), then mf (SQASS) ≥ 0.5*
4. *If (a=true), (b=false) and (c=false), then mf (NQASS) ≥ 0.25*
5. *If (a=false), (b=false) and (c=false), then mf (PSS) ≥ 0*

$$\mu_{(Subjective\ class)} = \begin{cases} 1 & p \ge (a+b+c) \\ 0.75 & c \ge p \ge (a+b) \\ 0.5 & b \ge p \ge (a+c) \\ 0.25 & (b+c) \ge p \ge a \\ 0 & (a+b+c) \ge p \end{cases} \tag{2.5}$$

Here, $(a+b+c) \Rightarrow (a\ \&\&\ b\ \&\&\ c)$ and $p \Rightarrow$ pattern.

Triangular membership function has been used to capture the boundaries for the patterns of *Subjective Active* class. The patterns *HQASS, QASS, SQASS* and *NQASS* of *Subjective Active* class are in the range of [0.25–1]. Linguistic grades (*Subjective Lower Medium, Medium, Higher Medium, High*) and boundaries between the patterns are captured successfully using POS variables of term features as shown in Fig. 2.5. For instance, **Rule:** If $s = VB$, $et = JJ/VB$, $ct_{(I)} \neq RB$ and $ct_{(NI)} = RB$, then class *SQASS* with *CF* is *Subjective Medium (SM)*.

Similarly in Fig. 2.6, the boundary levels for complete *Subjective* class patterns are depicted along with their linguistic grades. It is noticed that *Active* patterns [0.25–1] and *Passive* patterns [0–0.24] share the common membership scale [0–1].

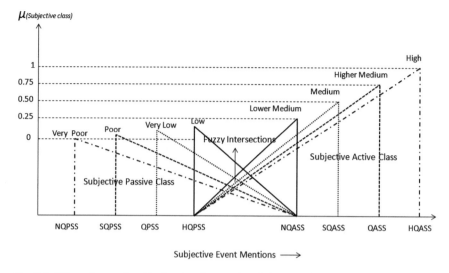

Fig. 2.6 Triangular membership function for complete *Subjective* class patterns

Similarly, overlapped patterns are obtained for complete *Objective* class in which both *Active* and *Passive* class patterns share common membership scale [0–1]. In this section, the importance of the patterns is estimated using the fuzzy membership values. As a result, each pattern is associated with a membership grade and the richness in the information content is identified. To make use of this idea, a weighting scheme for patterns is presented in the next section.

2.6.2   *Verification of Presented Fuzzy Rules Using Fuzzy Petri Nets (FPN)*

In general, anomalies in the rules are referred to as verification errors. They are inconsistency, incompleteness, redundancy, etc., and should be identified and removed for the error-free rules. Inconsistency in rules generates conflict results, say the rules R1: p1 ∧ p2 → p3, R2: p3 ∧ p4 → p5, R3: p1 ∧ p2 ∧ p4 → p5 are said to be inconsistent. Incompleteness in rules is generated by missing rules. For example, R1: p1 ∧ p2 → p3 is said to incomplete if p1 is a fact and p2 is neither a fact nor a conclusion of other rule. Consider rules R1: p1 ∧ p2 → p3, R2: p2 ∧ p3 and these rules are redundant, since they refer unnecessary rules in a rule base. Similarly, rules are said to be subsumed when two rules have identical conclusions. Consider the rules R1: p1 ∧ p2 → p3, R2: p2 → p3 where the antecedents of one rule are subset of the antecedents of another. Finally, if the rules have circular dependency, say R1: p1 → p2, R2: p2 → p3, R3: p3 → p1, it is said to be circular rule.

Based on the above discussion, it is imperative that the rule base constructed for any application domain should be validated for having a perfect rule set. In this section, Fuzzy Petri Nets (FPN) have been used for verifying the presented sentence classification model (SCM). The FPN is a combination of Petri Nets and rule-based knowledge representation. Discussing and explaining FPN is out of scope of this chapter, and interested reader can refer related research materials (He et al. 1999). The FPN for the presented SCM is introduced within a 5-tuple consisting of the Input Property Set (IPS), Internal Property Set (InPS), Output Property Set (OPS) and Rule Set (RS). Question 1 (Q1) to Question 13 (Q13) represent Input Properties such as term features (s, et, $ct_{(I)}$, $ct_{(NI)}$) and POS features (verb, noun, adjective, adverb) for the Internal Properties '$NQASS$', '$SQASS$', '$QASS$', '$HQASS$' patterns, respectively.

(i)   *The Input Properties*

The Input Properties are gathered within 13 questions and represented as Q1 to Q13 below:

Q1: is s is verb?	Q8: is $ct_{(I)}$ is adverb?
Q2: is s is non-verb?	Q9: is $ct_{(I)}$ is not adverb?
Q3: is et is noun?	Q10: is $ct_{(NI)}$ is adjective?
Q4: is et is adjective/verb?	Q11: is $ct_{(NI)}$ is not adjective?
Q5: is et is adjective?	Q12: is $ct_{(NI)}$ is adverb?
Q6: is $ct_{(I)}$ is adjective?	Q13: is $ct_{(NI)}$ is not adverb?
Q7: is $ct_{(I)}$ is not adjective?	

Though complete Input Properties set for all 16 rules of *Subjective* and *Objective* classes are given, for want of space and clarity, the rules are considered for the patterns of *Subjective Active* class (*ASS*) alone in further discussion and, however, can be extended for other class also.

(ii) *The Internal Properties*

The Internal Properties of the SCM are derived using various Input Properties as shown below:

1. The Input Properties Q1, Q4, Q9 and Q13 form an Internal Property called '*NQASS*' pattern.
2. The Input Properties Q1, Q4, Q9 and Q12 form an Internal Property called '*SQASS*' pattern.
3. The Input Properties Q1, Q4, Q8 and Q13 form an Internal Property called '*QASS*' pattern.
4. The Input Properties Q1, Q4, Q8 and Q12 form an Internal Property called '*HQASS*' pattern.

Below, Input and Internal Properties are deduced and presented in two levels:

(1) *Level 1:*

> If Q1 && Q4 && Q9 && Q13 exist, then '*NQASS*' pattern
> If Q1 && Q4 && Q9 && Q12 exist, then '*SQASS*' pattern
> If Q1 && Q4 && Q8 && Q13 exist, then '*QASS*' pattern
> If Q1 && Q4 && Q8 && Q12 exist, then '*HQASS*' pattern.

(2) *Level 2:*

If '*NQASS*' && '*SQASS*' && '*QASS*' && '*HQASS*' patterns exist, then sentence set belongs to '*ASS*' sub-class. Patterns are substituted with membership values. The sample rule base for the presented SCM is presented below in the structured format:

SCM = (Classification, IPS, InPS, OPS, RS)

SCM.IPS = {Q1, Q2, Q3, Q4, Q5, Q6, Q7, Q8, Q9, Q10, Q11, Q12, Q13}

SCM.InPS = {$NQASS$, $SQASS$, $QASS$, $HQASS$,}

SCM.OPS = {ASS}

SCM.RS = {R1, R2, R3, R4, R5}

SCM.RS.R1 = {Rule1, P1$\land$P3$\land$P9$\land$P13, P14, 0.25}

SCM.RS.R2 = {Rule2 P1$\land$P3$\land$P9$\land$P12, P15, 0.50}

SCM.RS.R3 = {Rule3, P1$\land$P3$\land$P8$\land$P13, P16, 0.75}

SCM.RS.R4 = {Rule4, P1$\land$P3$\land$P8$\land$P12, P17, 1}

SCM.RS.R5 = {Rule5, P14$\lor$P15$\lor$P16$\lor$P17, P18}

Verification Process

For verifying the rule base, it is mapped to a FPN as shown in Fig. 2.7. A reachability graph is generated using the ω-Net concept algorithm presented in He et al. (1999). The verification parameters, such as incompleteness, inconsistency, circularity and redundancy, are verified using the analysis presented in He et al. (1999).

Based on the obtained FPN, the reachability graph is presented in Fig. 2.8 and the following conclusions are drawn:

1. All the places (P) and transitions (T) exist, so there is no incompleteness errors.
2. P14, P15, P16 and P17 are different states of one property, so no inconsistency.
3. There is no loop in the reachability graph, and thus there is no circularity.
4. Finally, there is no redundancy since there is no transitions underlined.

It is observed that the fuzzy rules presented in this chapter (Table 2.6) is verified using FPN for identifying anomalies in the rule base. The reachability graph obtained using well-known algorithm found that there are no anomalies in the rule base.

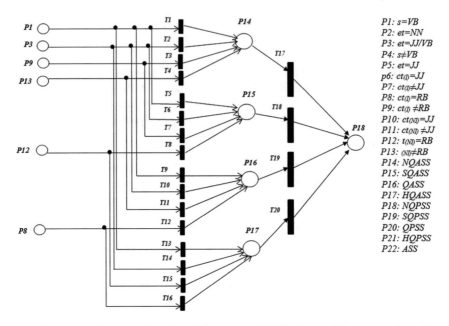

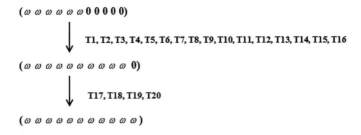

Fig. 2.7 Fuzzy Petri Net representation of the sentence classification model

Fig. 2.8 Reachability graph

2.7 Weights for Patterns Using Membership Function

It is noticed from the previous section that the patterns are differentiated based on the
information content captured by fuzzy rules and presented through linguistic grades.
The linguistic grades help in deriving the membership values which represents the
significance level of patterns. For instance, the membership scales of *Subjective
Active* class patterns are [0.25–1] and in the order of *HQASS, QASS, SQASS, NQASS*.
Similarly, the membership scale of *Subjective Passive* class patterns is [0–0.24] and
in the order of *HQPSS, QPSS, SQPSS, NQPSS*. This is represented in Fig. 2.9 in
a generic form in which *High-Qualified (HQ)*, *Qualified (Q)*, *Semi-Qualified (SQ)*
and *Non-Qualified (NQ)* patterns are represented based on their significant level (k)
order.

Fig. 2.9 Significant levels
for patterns

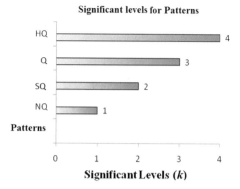

Based on the above significant level approach, the patterns are assigned weight. It is known that membership value of *Subjective* class is in the range of [0–1]; hence, the weight assigned for *Subjective* class is 1 (100%). The weight is further distributed for *Active* and *Passive* sub-classes. The sum of the weights of these sub-classes is equal to their parent class as represented in Eq. (2.6). Here, W_s represents the weight assigned to *Subjective* class, w_A and w_P are the weights assigned to *Active* and *Passive* sub-classes, respectively. The *Passive* sub-class weight w_P is one-fourth (0.24), since *mf(PSS)* ranges between [0–0.24] and *Active* sub-class weight w_s is three-fourth (0.75), since *mf(ASS)* ranges between [0.25–1].

$$W_S = (w_A + w_P) \tag{2.6}$$

Further, *Active* sub-class weight (w_A) is distributed to its patterns *HQASS*, *QASS*, *SQASS*, *NQASS*, i.e. $w_A = (w_{A1} + w_{A2} + w_{A3} + w_{A4})$ and the *Passive* sub-class weight (w_P) is distributed to its patterns *HQPSS*, *QPSS*, *SQPSS*, *NQPSS*, i.e. $w_P = (w_{P1} + w_{P2} + w_{P3} + w_{P4})$ as shown in Eq. (2.7). Here, w_{A1}, w_{A2}, w_{A3}, w_{A4} are weights assigned for the patterns (*HQASS*, *QASS*, *SQASS*, *NQASS*) of *Active* sub-class, and w_{P1}, w_{P2}, w_{P3}, w_{P4} are weights assigned for patterns (*HQPSS*, *QPSS*, *SQPSS*, *NQPSS*) of *Passive* sub-class.

$$w_A = \sum_{k=1}^{n} w_{Ak} \text{ and } w_P = \sum_{k=1}^{n} w_{Pk} \text{ where } n = 4 \tag{2.7}$$

The weights for these *Active* and *Passive* sub-class patterns are calculated using significant levels (k) of Fig. 2.9. This is depicted in Eq. (2.8).

$$w_{Ak} = \left[\left(\frac{k}{\sum_{k=1}^{n} k} \right) * w_A \right] \text{ and } w_{Pk} = \left[\left(\frac{k}{\sum_{k=1}^{n} k} \right) * w_P \right] \text{ where } n = 4 \tag{2.8}$$

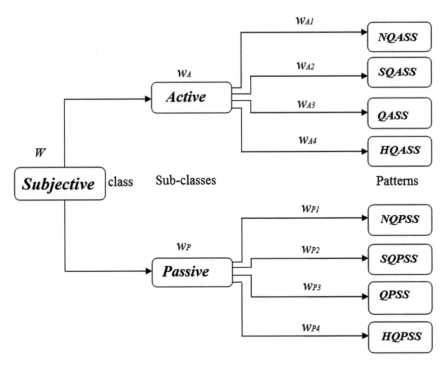

Fig. 2.10 Weights derived for *Subjective* class patterns

Figure 2.10 represents the derived weight hierarchy for *Active and Passive* sub-class patterns of *Subjective* class. In *Active* class, *HQASS* pattern is assigned higher weight w_{A4}, next weight w_{A3} is assigned for *QASS,* and lower w_{A1} weight is assigned to *NQASS*. Similar procedure can be followed for *Passive* class for assigning weights. The range of weights for various pattern are denoted in subscript value of weight, which range from maximum to minimum.

It is noticed that using fuzzy approach, rules are refined and more number of *Qualified* patterns (*HQASS, QASS, SQASS HQPSS, QPSS, SQPSS*) from *Subjective* class are extracted. The useful and interested information about Event happenings is present in these patterns. Similar process is used and analysis done for *Objective* class and extracted *Qualified* patterns (*HQAOS, QAOS, SQAOS, HQPOS, QPOS, SQPOS*). These patterns provide more insight knowledge about Specific Event Instance using named entity recognitions. Based on the information content expressed by the linguistic grades and weights estimated by membership function, the above-mentioned *Qualified* patterns of both *Subjective* and *Objective* classes are identified as Event Instance patterns and the information present in these Instances are used to construct the Event Corpus as depicted in Fig. 2.11. These patterns have proved the presence of useful information about Event Instance in the experiment section.

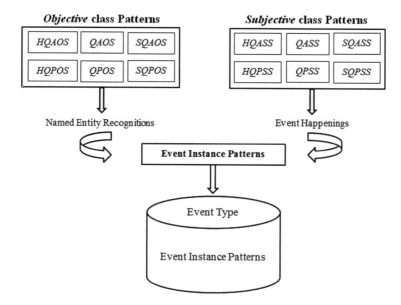

Fig. 2.11 Event Corpus built from Event Instance patterns

The constructed Corpus can be used for various application domains. For example, Internet search engine presents the retrieval set to the user query with URL along with three lines of description. The users, in general, go through the sentences for relevancy and choose the links accordingly. The presented Corpus can be used for this application to display the useful sentence patterns for a given Event Type (query). If the user's interest matches with the information provided by sentence patterns and needs detail description of the Instance, the related Web document can be chosen and retrieved. Apart from this, the presented approach can be useful similar to Wikipedia. It is well-known that Wikipedia is a great resource of heterogeneous topics. Based on user's interested topic, more knowledge can be gained by browsing topic-related hyperlinks. Apart from the crime domain, the presented approach Corpus can be built for various domains such as sports, entertainment and can be used to achieve more knowledge about the Events that are taking place in the respective domains.

2.8 Experimental Results

2.8.1 Performance Evaluation Using Controlled Dataset

For evaluating the presented approach, the standard Corpus is used as a collection of articles from the Iraq Body Count (IBC) database (http://www.iraqbodycount. org/database/). The database is manually annotated with the 'Die' Event Type as

Table 2.7 Sample Event Types and their Trigger terms

Event Type	Trigger terms
Die	15
Kidnap	08
Kill	11
Wound	09
Shoot	07

Table 2.8 IBC Corpus Statistics for 'Die' Event Type

Features	'Die' Event Type
Number of documents	332
Number of Event sets	101
Number of sentences	8628
Number of Event-related sentences (without synonyms)	358
Number of sentences in Objective class	223
Number of sentences in Subjective class	135

mentioned in Naughton et al. (2010). This dataset is developed from larger research projects that mainly focused on statistical approaches for collecting fatality statistics from unstructured news data during the Iraq War. Various Event Types such as Die, Kill, Torture, Shoot, Wound and Attack have been used in the experiments. Table 2.7 shows the sample Event Types and Trigger terms that are taken for processing. For instance, {suicide, expires, break, etc.,} are the synonyms/hyponyms/hypernyms that exist for 'Die' Event Type.

Since IBC Corpus deals with the death incident, evaluation is initiated by using this Corpus for classifying the Event Mention patterns based on 'Die' Event Type. The Corpus features of 'Die' Event Type are presented in Table 2.8 and have 332 documents from 77 sources. The collections of articles are categorized into 101 Event sets, such that documents that have common Events are grouped into the same set. The annotation for 101 Events is done manually, which is a tedious and vital process. During this process, the sentences are identified and annotated to get description about 'Die' Event. The total number of sentences obtained from 101 Events is 8628. These gold standard annotations are collected to evaluate the usefulness of the presented approach. Initially, a small sample set with 358 sentences for Event Type 'Die', without using synonyms, hyponyms and hypernyms, is considered to generate annotations. Based on the $s = VB$ feature, the sentences are grouped into *Objective* and *Subjective* classes and are shown in Table 2.8.

Initially, the *Subjective* and *Objective* classes are manually annotated by making different groups. The process of annotation is carried out by two groups of undergraduate students based on verb POS nature of sentence using NLP tool. Classification based on *et* terms is also carried out for Event Type 'Die'. The classification based on the *et* terms gives *Active* and *Passive* sub-classes of *Subjective* and *Objective* classes.

The final-level classification is based on the $ct_{(I)}$ and $ct_{(NI)}$ terms of et in the sentence which gives *Qualified* and *Non-Qualified* patterns of *Subjective* and *Objective* classes. All the classification output, based on annotation, is re-evaluated by a group of research students for obtaining eight patterns for each class. The ANFIS model is used as classifier with Sugeno-type FIS structure for training the data. The back propagation technique is used for training the FIS structure. For an *Subjective* and *Objective* sentence set, three term features such as et, $ct_{(I)}$ and $ct_{(NI)}$ are given as input and sixteen rules are obtained for the classifying the patterns. In *Subjective* class, the sentences are classified into eight patterns and in *Objective* class, the sentences are classified into eight patterns. As performance measure classification accuracy is used, which is defined as the ratio of sentences correctly classified by the classifier to class type n to the human-annotated sentences for the class type n? This measure estimates the classification accuracy done by fuzzy rules for sentences that are classified, and the results are shown in Table 2.9. The misclassified sentences are 24% and represented as outliers for *Objective* class and 18% for *Subjective* class. This may be due to the unrecognized Event Types in the sentence, which conflicts and misleads the classifier during classification. It is noticed that the difference between classifier and manual annotation is less, i.e. in the range of 2-7% for each pattern. Also, the classification accuracy between ANFIS classifier and human annotation matches above 75% for a sample dataset.

Further, synonyms, hyponyms and hypernyms of 'Die' Event are considered, and 2332 sentences are extracted from respective documents, out of which 1376 sentences are classified as *Objective* class and 956 sentences are classified as *Subjective* class. The sentences of *Objective* and *Subjective* classes are further classified into sixteen patterns as shown in Table 2.10. The sixteen patterns obtained are used to analyse the performance of rules by classifier using k-fold cross-validation technique. The validation process is initiated by considering twofold cross-validation (holdout method) as it consumes less computational time. The sentences are randomized using random generator and chosen as equal-sized validation and training dataset. The twofold cross-validation is performed using training set and expects the classifier to predict the output values for the testing dataset, where these output values have no prior appearance. The performance of classification for this iteration is considered in terms of precision and recall. In the next iteration, again the sentences are randomized with different validation and training datasets, cross-validation process is repeated and evaluated. Similarly, the sentences are randomized for n iterations ($n = 100$), and the average classification performance is considered and evaluated. The average results obtained after n iterations are good. The advantage of iterative twofold cross-validation is that it matters less how the data gets divided since every data point is randomized and appears at least once either in a test set or training set in each iteration. Here, the variation between training and test datasets is reduced as the iteration is increased. Later, tenfold cross-validation is used to evaluate the performance of rules by dividing the dataset into ten sets and the cross-validation is repeated 10 times. Each time, one of the ten subsets is used as the test set and the other nine subsets are put together to form a training set. The 10 results from the folds are averaged to produce a single estimation. The well-known performance measures

Table 2.9 Classification accuracy of human annotation and ANFIS for Event Type 'Die'

Manual Annotation		ANFIS rules		
Objective class patterns	Patterns count (%)	Objective class patterns	Patterns count (%)	Classification accuracy (%)
NQAOS	8.17	NQAOS	6.4	73.56
SQAOS	17.04	SQAOS	12.03	70.5
QAOS	3.5	QAOS	3.0	85.7
HQAOS	15.69	HQAOS	12.78	81.5
NQPOS	13.45	NQPOS	10.34	76.8
SQPOS	19.73	SQPOS	15.22	77.1
QPOS	7.17	QPOS	6.12	81.4
HQPOS	14.34	HQPOS	10.09	70.36
		Outliers	24.02	
Subjective class patterns	Patterns count (%)	Subjective class patterns	Patterns count (%)	Classification accuracy (%)
NQASS	5.44	NQASS	3.2	58.8
SQASS	12.85	SQASS	10.11	78.6
QASS	5.0	QASS	4.2	84.0
HQASS	19.74	HQASS	16.2	82.0
NQPSS	10.07	NQPSS	9.89	98.2
SQPSS	20.29	SQPSS	16.34	80.5
QPSS	4.01	QPSS	3.0	75.1
HQPSS	22.21	HQPSS	19.4	87.3
		Outliers	18.00	

such as Precision, Recall and F1 measure are used for evaluation. Precision is defined as the ratio of number of sentences correctly classified by a classifier to a class type n to the total number of sentences classified by a system to a class type n. Recall is defined as the ratio of number of sentences correctly classified by a classifier to a class type n to the total number of human-annotated sentences of class type n, and F1 measure gives the harmonic measure of Precision and Recall for class type n. The results are shown in Table 2.10.

It is noticed from Table 2.10 that the precision, recall and F1 measure for both *Subjective* and *Objective* classes are good. The classified patterns for *Objective* class are more compared to *Subjective* class patterns. This is due to the fact that the number of patterns in *Objective* class is more and the classifier acquired more intuition. As a result, only higher weighted *Qualified* patterns belonging to *Objective* and *Subjective* class are considered, since they provide useful and interested information about Instances related to 'Die' Event. Thus, high-weighted *HQ, Q, SQ (HQAOS, QAOS, SQAOS, HQPOS, QPOS, SQPOS, HQASS, QASS, SQASS, HQPSS, QPSS, SQPSS)* Event Instance patterns are considered from both *Objective* and *Subjec-*

Table 2.10 Performance of ANFIS using tenfold cross-validation for Event Type 'Die'

Number of Event-related sentences (with synonyms): 2332
Number of Subjective sentences: 1376
Number of Objective sentences: 956

Objective class patterns	Prec (%)	Recall (%)	F1 (%)	Subjective class patterns	Prec (%)	Recall (%)	F1 (%)
NQAOS	98.11	93.1	96.23	NQAOS	87.2	82.34	86.45
SQAOS	95.02	92.45	93.2	SQAOS	89.87	86.7	86.04
QAOS	98.45	90.2	94.8	QAOS	85.44	84.2	84.9
HQAOS	94.65	95.23	94	HQAOS	94.9	93.1	93.02
NQPOS	96	90.02	93.4	NQPOS	82.07	80.09	81.71
SQPOS	94.76	94	94.4	SQPOS	91	93.65	92.5
QPOS	98.11	95.12	97.1	QPOS	89.06	95.23	90.03
HQPOS	93.01	97.01	95.8	HQPOS	90.31	91.19	90.74

Table 2.11 Performance evaluations (%) for IBC dataset using k-fold cross-validation

Event Type 'Die'	Twofold cross-validation			Tenfold cross-validation		
	Prec	Recall	F1	Prec	Recall	F1
Objective class Instance patterns	94.54	90.25	92.34	95.66	94.01	94.88
Subjective class Instance patterns	88.49	86.33	87.39	90.09	90.67	90.37

tive class. The precision, recall and F1 measures are averaged for the patterns that belong to *Objective* class and *Subjective* class and presented in Table 2.11. The performance of the presented approach is also evaluated using twofold cross-validation. The average results of both twofold and tenfold cross-validation are compared. Both cross-validations are performed to ensure the classification performances. The average results in both the cases are good with negligible variation of 3-4%. This is common scenario from all the classifiers, since the evaluation depends heavily on the data points that fall in the training and the testing set.

To further consolidate the performance of the presented approach, the comparison is done with the other similar Event detection approaches (Naughton et al. 2010) that used SVM classifier and Trigger-based classifications. In the SVM-based classification approach, the classifier uses terms, lexical and additional Event-based features to encode each training/test Instance and classify Event Instances in binary mode of On-Event and Off-Event class. The Trigger-based approach of Event detection system uses NLP approach by developing a manual rule-based system, which finds sentences connected to a target Event Type using a handcrafted list of trigger terms found in WordNet. In the presented approach, the average results of high-weighted *Qualified* Event Instance patterns (*HQ, Q, SQ*) from both classes are considered and compared with SVM- and Trigger-based approach. Table 2.12 presents the performance results

Table 2.12 Performance measure (%) of presented approach with other approaches for Event Type 'Die'

Approaches	Twofold cross-validation			Tenfold cross-validation		
	Prec	Recall	F1	Prec	Recall	F1
Presented approach	91.51	88.29	89.86	92.87	92.34	92.62
SVM approach	88.00	86.9	87.44	90.00	91.9	90.94
Trigger based	79.3	80.5	79.79	83.3	92.5	87.65

of the presented approach with the comparative approaches for an Event Type 'Die'. The results are evaluated for both twofold and tenfold cross-validation.

The presented approach achieved better results compared to comparative approaches. This is due to the fact that the presented approach refines the rules in many levels and classifies the sentences into many patterns based on the type and useful content of the information. In contrast, SVM classifier uses the binary classification (On-Event and Off-Event), where 90% sentences constitute for Off-Event class.

In Fig. 2.12, only F1 score is presented, since it is the harmonic means of precision and recall and it is felt that at this point it is enough to present F1 score alone. Figure 2.12 (a) depicts the F1 scores for all the approaches using varying levels of training data for Event Type 'Die'. These scores are averaged over 5 runs using 50:50 random training and testing splits for 'Die' Event. The SVM approach and trigger-based approach obtain F1 scores 86% and 82% using 10% training data. It is observed from the result that F1 score of the presented approach is around 60% when 10% data is used for training. The performance increases efficiently with the size of the training data.

While the training dataset is more than 50%, the presented approach achieves marginally higher F1 scores for both the classes compared to other methods. The experimentation is done by considering other Event Types such as 'Wound' and 'Attack', apart from 'Die' Event Type and evaluated the performance and compared with SVM- and Trigger-based approaches. The SVM has used features such as SVM(terms), SVM(terms + nc) and SVM(terms + nc + lex), which considers terms, noun chunks(nc) and lexical information such as POS tag and chunk tags(lex). The result is presented in Fig. 2.12 (b) and observed that the performance of the presented scheme is encouraging.

2.8.2   Performance Evaluation for Uncontrolled Dataset Generated from WWW

The evaluation on the performance of the presented approach using Corpus generated from Web documents (Web Corpus) by crawling crime-related Web pages from http://www.trutv.com/library/crime/ is done.

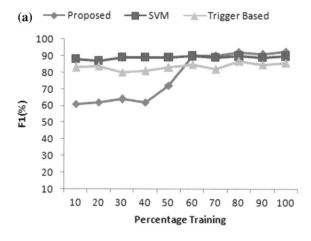

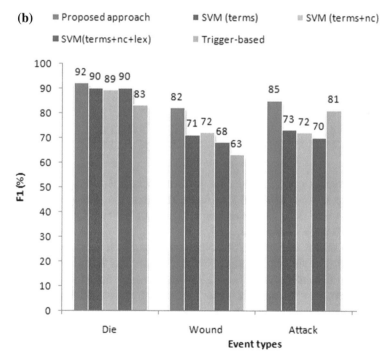

Fig. 2.12 F1 score (%). **a** Various training data and **b** various Event Types

The HTML pages are parsed; text content is extracted and preprocessed by removing stop words. The features of Web Corpus are shown in Table 2.13. From the crime library articles, 500 Web documents are collected in which 500 *et* terms are chosen and 46,938 sentences are obtained. Among them, *Subjective sentences* are 14,675

Table 2.13 Web Corpus Statistics from www.trutv.com/library/crime

Corpus features	Homogeneous Corpus
Number of documents	500
Number of et terms	500
Number of sentences	46,938
Objective class sentences	32,263
Subjective class sentences	14,675
Event Types for analysis	4
Objective class for selected Event Types	6613
Subjective class for selected Event Types	4211

Table 2.14 F1 measure of the presented approach with other approaches for various Event Types in Web Corpus

Event Types	F1(%) for various approaches				
	Presented approach	SVM classification approach			Trigger based
		SVM (terms)	SVM (terms + nc)	SVM (terms + nc + lex)	
Die	94.2	73.0	76.22	90.91	73.65
Wound	89.42	71.84	72.11	81.2	63.6
Attack	95.9	73.65	72.84	86.14	81.5
Kill	85.1	78.76	82.14	85.32	70.7

and *Objective sentences* are 32,263. For the analysis, four Event Types such as 'Die', 'Wound', 'Attack' and 'Kill' and their Event Triggered terms have been considered for annotation. For these Triggered terms, the *Objective* and *Subjective* sentences obtained are shown in Table 2.13.

Table 2.14 shows the F1 score of the presented approach for various Event Types and compared with various features of SVM classifier and Trigger-based approach. The ground truth is measured by human annotation as it is done in earlier cases. The F1 score differs for various Event Types, since the Events in the Web Corpus have diverse contexts and topics. It is noticed that the fuzzy rule achieves good results in classifying the Event Mention sentences for identifying the interested information about Event Instances even for uncontrolled dataset (Web Corpus). Based on the experimental results, it is imperative that the presented approach is useful for domain-specific Web applications compared to other approaches.

Table 2.15 F1 measure for various combinations of training/test data for the 'Die' Event

Training/Testing datasets	F1(%) for various approaches		
	Presented approach	SVM (terms + nc + lex) approach	Trigger based
Train:Web/IBC versus Test:Web/IBC	89.32	82.03	75.1
Train:Web versus Test:Web	94.2	84.33	78.89
Train:IBC versus Test:IBC	92.62	86.01	76.01
Train:Web versus Test:IBC	87.64	78.45	69.5
Train:IBC versus Test:Web	86.13	79.08	65.65

2.8.3 Web Corpus Versus IBC Corpus

Based on all the performance analysis, it is found that information content in the Event Instance patterns that are classified by using a fuzzy approach is useful and can be used for any application domain. Thus, all the high-weighted *Qualified* patterns (*SQ, Q, HQ*) that belongs to *Subjective* and *Objective* classes are considered for an Event Type 'Die', and are updated into the Corpus. The *IBC* and *Web* Corpus are newly created for an Event Type 'Die' with various Event Instance patterns and evaluated using different combination of testing and training dataset. When IBC and Web Corpus datasets are compared for the 'Die' Event, some of the sentence properties change with the topic of interest and the size, which influences the performance. The behaviour of the presented approach is evaluated for the 'Die' Event using the following combinations of training/test data and compared F1 scores using tenfold cross-validation as shown in Table 2.15.

Train:Web/IBC versus Test:Web/IBC—Mixture of IBC and Web datasets is used for both training/testing.
Train:Web versus Test:Web—Web dataset is used for both training and testing.
Train:IBC versus Test:IBC—IBC dataset is used for both training and testing.
Train:Web versus Test:IBC—Web data is used for training, and IBC data is used for testing.
Train:IBC versus Test:Web—IBC data is used for training, and Web data is used for testing.

Table 2.15 presents the sensitive changes in F1 score by various approaches. This is due to the fact that Event features are mutually dependent on each other starting from the seed *et* terms to the Event Type. Initial variations in the seed triggered and sentence classification effects the Event Instances patterns further. There are variations in F1 score for every combination of training and testing dataset of Web and IBC Corpus. In general, classification is good in the heterogeneous dataset due to the vast variations in the information content of seed terms. However, the objective is to classify the sentences in the homogeneous dataset, where the information content of seed terms is interrelated to each other in either way. In addition, most of the systems find it more difficult to classify Event Instances patterns accurately that contain Events described across more diverse contexts, topics and circumstances. Classifying the Event patterns for Event Instances in the homogeneous dataset is difficult and a challenging task is performed by the presented approach and is compared again with SVM- and Trigger-based approach. The combination of *Train:IBC versus Test:IBC*, *Train:Web versus Test:Web* and *Train:IBC/Web versus Test:IBC/Web* Corpus dataset produces good results across other training and testing combinations. This is due to the fact that the training and testing dataset Corpus is same and the presented approach effectively classifies the Event Instance patterns in refined manner though they have more diverse contexts, topics and circumstances.

The combination of *Train:IBC versus Test:Web* and *Train:Web versus Test:IBC* Corpus dataset produces low results since, content-wise, both Corpus size and topics vary. This is due to the presence of information content errors and errors in seed triggers; the systems find it more difficult to correctly classify the Event Instance patterns, especially when they are described across more diverse contexts, topics and circumstances. Even though these limitations are true for all the approaches that are used for evaluation, the presented fuzzy approach has achieved good results, comparatively, when a different Corpus is used as training/test set. For completeness, the constructed Corpus is evaluated for the amount of interested information present in terms of Event Instances for an Event Type. If I_R is the total number of Instances correctly classified (True Positives and True Negatives) for an Event Type and I_T is the total number of human-annotated Instances in the collection, then the classification accuracy is defined as shown in Eq. (2.9).

$$\text{Accuracy} = \frac{I_R}{I_T} \tag{2.9}$$

The presented approach is evaluated for various Event Types using the constructed Corpus in identifying related Event Instances and compared with other similar approaches and is noticed that the presented approach achieves higher accuracy compared to other approaches. The presented approach is useful in providing the interested information by building the Event Corpus, and the result is shown in Table 2.16.

Based on all the performance evaluation, it is observed that the classification accuracy achieved by the presented fuzzy rules is good. While Corpus is generated using the patterns, it is useful for obtaining interested and useful information. For domain-specific applications, the Corpus content can be effectively used for obtaining useful information related to domain-related queries. The generated Corpus can be used in information retrieval application for retrieving relevant information with respect to a domain for further analysis and prediction. The retrieval set can be ranked using the weights assigned in Sect. 2.7.

Table 2.16 Accuracy for the Instances for various Event Types

Event Types	Accuracy(%) for various approaches		
	Presented approach	SVM (terms + nc + lex)	Trigger based
Die	93.2	89.43	84.0
Wound	78.65	82.03	75.66
Attack	91.23	87.76	90.02
Kill	89.07	78.52	82.34

2.9 Conclusion and Future Works

The chapter presented a fuzzy-based approach for Event detection using sentence features. Based on the POS nature of the sentence terms, a hierarchical classification model is presented and eight patterns have been obtained. The CART is used for validating the model, and it is found that the boundaries between the classes are not well defined. Hence, fuzzy rules are generated and sixteen patterns are obtained. Each pattern is denoted with appropriate linguistic grade to represent interested and useful information content about the Instance for a given Event Type. The rules are evaluated using suitable membership function by assigning membership values to patterns. Next, suitable weights are assigned based on the pattern significances. Higher weighted Instance patterns are considered to build the Corpus and used for domain-specific retrieval applications. The IBC benchmark dataset and Corpus generated from Web document are used for evaluation. Many Event Types such as 'Die' and 'Kill' have been considered and evaluated the performance. The classification accuracy is encouraging compared to some of the other similar approaches. As a future work, the presented approach can be used for building the Corpus with various Event Types such as sports and entertainment and can be used for domain-specific retrieval application.

References

Abuleil, S. (2007). Using NLP techniques for tagging events in Arabic text. In *Proceedings of 19th IEEE International Conference on Tools with AI* (pp. 440–443). New York: IEEE Press.

ACE (Automatic Content Extraction) English Annotation Guidelines for events Version 5.4.3 2005.07.01 Linguistic Data Consortium. http://www.ldc.upenn.edu.

Allan, J., Jaime, C., George, D., Jonathon, Y., & Yiming, Y. (1998). Topic detection and tracking pilot study.

Allan, J., Wade, C., & Bolivar, A. (2003). Retrieval and novelty detection at the sentence level. In: *Proceedings of SIGIR* (pp. 314–321).

Aone, C., & Ramos-Santacruz, M. (2000). REES: A large-scale relation and event extraction system. In: *Proceedings of 6th Conference on Applied Natural Language Processing* (pp. 76–83). Washington: Morgan Kaufmann Publishers Inc.

Breiman, L., Friedman, J., Olshen, R. A., & Charles, J. S. (1985). *Classification and regression trees*. Monterey, CA: Wadsworth and Brooks.

Creswell, C., Beal, J. M., Chen, J., Cornell, L. T., Nilsson, L., Srihari, R. K. (2006). Automatically extracting nominal mentions of events with a bootstrapped probabilistic classifier. In *Proceedings of COLING/ACL 2006 Main Conference Poster Sessions* (pp. 168–175).

Cohen, K. B., Verspoor, K., Johnson, H., Roeder, C., Ogren, P., Baumgartner, W., et al. (2009). High-precision biological event extraction with a concept recognizer. In *Proceedings of BioNLP09 Shared Task Workshop* (pp. 50–58).

David, A. (2006) Stages of event extraction. In *Proceedings of COLING/ACL 2006 Workshop on Annotating and Reasoning about Time and Events* (pp. 1–8).

Filatova, E., & Hatzivassiloglou, V. (2004). Event-based extractive summarization. In *Proceedings of ACL 2004 Workshop on Summarization*, Barcelona, Spain (pp. 104–111).

Grishman, R., & Sundheim, B. (1996). Message understanding conference-6: A brief history. In *Proceedings of Computational Linguistics*.

He, X., Chu, W. C., Yang, H., & Yang, S. J. H. (1999). A new approach to verify rule-based systems using petri nets. In *Proceedings of International 23rd IEEE Conference on Computer Software and Applications Conference* (COMPSAC'99) (pp. 462–467).

http://www.iraqbodycount.org/database/.

http://www.trutv.com/library/crime/.

Hristidis, V., Valdivia, O., Vlachos, M., & Yu, P. S. (2006). Continuous term search on multiple text streams. In *Proceedings of CIKM*, Arlington, VA (pp. 802–803).

Hung, S. H., Lin, C. H., & Hong, J. S. (2010). Web mining for event-based commonsense knowledge using lexico-syntactic pattern matching and semantic role labeling. *Journal of Expert Systems with Applications, 37*, 341–347.

Jang, J. S. R. (1993). ANFIS: Adaptive-Network-Based Fuzzy Inference System. *IEEE Transactions on Systems, Man, and Cybernetics, 23*(No. 5, Issue 6), 665–685.

Makoto, M., Rune, S., Jjin-Dong, K., & Junichi, T. (2010). Event extraction with complex event classification using rich features. *Journal of Bioinformatics and Computational Biology, 8*(1), 131–146.

McCracken, N., Ozgencil, N. E., & Symonenko, S. (2006). Combining techniques for event extraction in summary reports. In *Proceedings of AAAI 2006 Workshop Event Extraction and Synthesis* (pp. 7–11).

Naughton, M., Stokes, N., & Carthy, J. (2010). Sentence-level event classification in unstructured texts. *Information Retrieval, 13,* 132–156.

Xu, F., Uszkoreit, H., & Li, H. (2006). Automatic event and relation detection with seeds of varying complexity. In *Proceedings of AAAI 2006 Workshop Event Extraction and Synthesis*, Boston (pp. 491–498).

Yang, Y., Pierce, T., & Carbonell, J. (1998). A study on retrospective and on-line event detection. In *Proceedings of SIGIR* (pp. 28–36).

Yang, Y., Zhang, J., Carbonell, J., & Jin, C. (2002). Topic-conditioned novelty detection. In: *Proceedings of SIGKDD* (pp. 688–693).

Zhao, Q., Bhowmick, S. S., & Sun, A. (2006). *i*Wed: An integrated multigraph cut-based approach for detecting Events from a website. In *Proceedings of 10th Pacific-Asia conference on Advances in Knowledge Discovery and Data Mining* (pp. 351–360). Berlin, Heidelberg: Springer.

Zhao, Q., Liu, T. Y., Bhowmick, S. S., Ma, W.-Y. (2006). Event detection from evolution of click-through data. In *Proceedings of 12th ACM SIGKDD International Conference on Knowledge Discovery and Data Mining* (pp. 484–493). New York, NY, USA: ACM.

Zhou, W., Liu, Z., & Kong, Q. (2008). A survey of Event-based knowledge processing. *Chinese Journal of Computer Science, 33*(2), 160–162. (in Chinese).

Chapter 3
Constructing Thesaurus Using TAG Term Weight for Query Expansion in Information Retrieval Application

3.1 Introduction

The information retrieval systems are used by many people for searching relevant documents from WWW. Often, the query is short, vague and ambiguous. As a result, the search engine systems face difficulties in understanding these kinds of queries. Consider a query having text as 'Apple Price Today'. The ambiguity is inherently present in language by which the retrieval performance is low. Sometimes, while a well-formulated query string is present, similar term/text may not be appearing in web documents. This situation is also lower down the precision of retrieval. Thus, the precision of retrieval can be improved by effectively matching the content/pattern present in web documents and the query pattern. From the user's perspective, formulating effective query term is a challenging task as the contextual meaning of the query changes. The query refinement and query expansion are very popular schemes to handle the above-mentioned difficulties. These techniques are used by search engine systems for improving the precision of retrieval. Cucerzan and Brill (2004) have statistically presented that search engine has to use query refinement mechanism by 50%. There has been research in this domain for a long time. Most of the research group uses Web as Corpus for testing the procedure and algorithms. The content of Web as Corpus is multimodal and consists of e-texts. The researchers can create Corpora by using web derives for expanding ad refining the query (Marianne et al. 2007). In general, the query log is used by the retrieval system to determine web count. The retrieval system provides the same query sequence to predict the query pattern (Kilgarriff 2007). This process degrades the precision of retrieval. All the above-mentioned issues have been handled by Thesaurus for expanding queries. It is constructed using the content from inverted index having terms, document link or ID, along with weight of the term. The weight of each term is calculated based on the

© Springer Nature Singapore Pte Ltd. 2018
S. G. Shaila and A. Vadivel, *Textual and Visual Information Retrieval using Query Refinement and Pattern Analysis*, https://doi.org/10.1007/978-981-13-2559-5_3

properties of HTML TAGs and its frequency of occurrence. The weight assigned to the term is referred to as TAG term weight (TTW). Using the information available in the inverted index, the Thesaurus is constructed, say N-gram Thesaurus. As the N-gram Thesaurus is constructed only using the terms in HTML documents, it is a suitable candidate for search engine systems. The TREC recommended datasets have also been used for evaluating the performance of the N-gram Thesaurus. The precision, mean average precision (MAP) and mean reciprocal recall (MRR) are used as evaluation parameters. While comparing the performance of N-gram Thesaurus, it is observed that it has outperformed some of the well-known schemes.

The rest of this chapter is organized as follows. Section 3.2 presents related works, and N-gram Thesaurus is explained in Sects. 3.3, 3.4, 3.5, and algorithm is presented in 3.6. In Sect. 3.7, the experimental result is presented and concluded in the last section of the chapter.

3.2 Reviews on Query Expansion and Refinement Techniques

This section presents the literature review of the recently proposed approaches on query expansion along with term weighting methods. The query expansion methods may be classified into two categories such as global and local (Manning et al. 2008). The semantic relevance of the individual query terms is considered for reformulating the query, where the WordNet or Thesaurus is used. In contrast, the retrieved document for the original query is used for reformulating the query is called as local approaches. Suitable feedback mechanisms, say, relevance feedback, are used for refining the query term. The assumption is that the top-ranked documents are found relevant to the original query and contains relevant terms. These relevant terms are used for reformulating the query terms. However, it may not be always possible to retrieve relevant documents on the top of the retrieval set. This is true for the query with difficult and ambiguous terms. As a result, the query expansion and reformulation of query by these categories of approaches fail in terms of topic. Thus, N-gram Thesaurus is proposed in this chapter following global approach for constructing the knowledge source and the review in this section is also presented on global approaches.

There are approaches, which automatically use the Corpus for constructing Thesaurus. One of the well-known Corpus systems is inferred from Unified Medical Language (UML). Each concept of the canonical term is effectively used for building Thesaurus, and query expansion is performed. The query term is expanded using its synonyms and other related terms. In this approach, the expanded query term is assigned lower weight compared to the original query term. Yet another method has been proposed for Thesaurus construction. The Thesaurus is constructed manually only with synonyms of the concepts, and example of this kind of Thesaurus is UML meta-Thesaurus. The co-occurrence terms or grammatically related terms are used for designing other kinds of Thesaurus and are refered as automatically derived Thesaurus. The grammatical dependencies and frequent co-occurring terms are co-related for finding the grammatical dependency. The very popular WordNet expands the query by including the synonyms of the query terms (Voorhees 1994). There has been a difference between the retrieved set of original and refined query. Smeaton et al. (1995) have proposed a mechanism, which uses WordNet to expand the query. The hierarchical terms of the tree-based query are processed and expanded, and each term in the hierarchy is expanded with equal importance. However, it is very difficult to trace back the original query term. A query term and corresponding similarity matrix have been derived using index structure of query (Qiu and Frei 1993). This system performs well only for lesser number of terms and fails considerably with higher number of terms.

Jing and Croft (1994) have constructed Thesaurus by analysing the text and feature of the text. This approach has used phrase funder program that automates the entire procedure. However, the performance is very marginal. A Thesaurus has been constructed based on co-occurrence of words with the help of WordNet and Term Semantic Network (Gong et al. 2005). The Term Semantic Network is a filter and complemented the operation of WordNet. It is observed that the Thesaurus construction strategy has notable flaws. The Wikipedia is well-known Corpus and is available free of cost for the research community. As a result, most of the methods have used the same for expanding the query. It is a structured source of knowledge having information on varied topics and found to be suitable for query expansion. Different categories of articles in Wikipedia Corpus are assigned proper labelling for expanding query (Li et al. 2007). Each category is assigned weight, and it is used for raking it. However, while week queries are presented in the query interface, there is only little improvement in the performance.

Milne et al. (2007) have used Wikipedia and applied Thesaurus-based expansion scheme. The content of the Thesaurus is in line with the topics and relevant to the document collection. The topics of relevancy of the term with the topic are presented by extracting the consecutive sentence, which is related to the topic in the Thesaurus. The queries are grouped as entity, ambiguous and broader queries with the help of information content of Wikipedia (Xu et al. 2009). The top-ranked articles and entity pages from Wikipedia are used for retrieving pseudo-relevant text.

Kaptein and Kamps (2009) have proposed an approach, where the class information is used for expanding the user query. The distance between target and query category is calculated with a suitable weight function. However, there is a mismatch in the information content in category-related information. The candidate term of the original query term is extracted using co-occurrence approach (Van Rijsbergen

1977). The influence of the noise term and stop words has degraded the performance of this approach. The probability of query terms present in collection is analysed and extracted by Kullback–Liebler divergence (KLD) (Cover and Thomas 1991) and Bose–Einstein statistics weighting model (Macdonald et al. 2005). The probability co-occurrence term is also calculated, which in turn degrades query expansion. Perez-Aguera and Lourdes-Araujo (2008) have combined both the methods mentioned above and found the performance of the considered approach is very marginal compared to individual method.

In addition to the methods for constructing Thesaurus, assigning suitable weights in Thesaurus is also very important. The query term and terms present in documents are assigned suitable weights. The correspondence between these terms is captured in terms of weights. These weighting modes are used for ranking, say, automatic query concept, concept weighting. Vector space model (VSM) has been proposed, where both the query and documents are represented as vector (Lin et al. 2005). A set of fuzzy rules are applied to find the additional query terms. The importance of those additional terms and its relevancy are also calculated for assigning the weights. The association rule mining concept is applied for refining the query (Martin-Bautista et al. 2004). The association rules are applied on web documents for estimating the text transaction. The weight of the query terms is reassigned to mine more query terms (Wang et al. 2010). The similarity between the query terms and all the terms available in the document is calculated for reweighting the query term. A bipartite graph is constructed with query term as nodes and click as edges. The URL is labelled on the edges, and the weight of the edge is number of clicks.

The expanded query is formed by applying random walk probability. However, the performance of this approach over a baseline is very low. The query and document are a structured form and named as weight word pairs (WWP) (Francesco et al. 2013). An explicit feedback mechanism is formulated for representing the pair. In addition to the above methods, there are, say, latent Dirichlet allocation model (Blei et al. 2003) and probabilistic topic model (Griffiths et al. 2007). However, these methods expand the query by using only the indigenous knowledge. Metzler and Croft (2007) have proposed latent concept expansion (LCE), where the query is expanded using the concept of term dependency. The weight of the query term is calculated based on both syntactic and semantic dependencies. It is found that the above-mentioned dependency is orthogonal in nature. The query term ad query phrases are assigned equal and constant weights. However, the performance is very low.

Based on the above review, the global Thesaurus construction approach is suitable for retrieval applications. Further, appropriate weighting model is required to assign suitable weights to the terms in web documents. As a result, a scheme is presented for assigning weights to the terms in documents. These terms are updated in the Thesaurus through inverted index. Finally, the syntactic context of the terms, synonyms, etc., is found out for effectively constructing N-gram Thesaurus. The performance of N-gram Thesaurus with some of the other approaches is compared and found that the performance of the presented approach is encouraging.

3.3 Architecture View of Query Expansion

Most of the times, the query submitted by the user in the search interface of the retrieval system is very short. These short queries may not reflect the exact need of the information content from WWW and considered as baseline for initiating the initial retrieval. The retrieval set is refined by reformulating the initial query, where the query term is refined by having relevant keywords. The objective of the proposed N-gram Thesaurus is to expand the query suitably. Information retrieval applications fetch HTML documents from the Internet. These documents are processed automatically without the intervention for complementing the query expansion procedure. The N-gram Thesaurus is updated periodically with document information/document id, terms(s) and corresponding weight of the term. The functional units of the proposed N-gram Thesaurus are depicted in Fig. 3.1. There are two main functional categories such as TAG term weight (TTW) and query expansion. In the first stage, HTML documents are processed for obtaining the clear text by tokenizing the texts, stemming the text, etc. The final version of HTML text is updated in the inverted index along with other important information such as TAG weight (tg), frequency of occurrences (f) and the document id's (d).

In the second stage, the Unigram Thesaurus is updated with the content of inverted index. At a later stage, the co-occurrence terms of Unigram along with weight and document id are updated in N-gram Thesaurus accordingly. During the retrieval process, the query is considered as N-gram and N+1 gram Thesaurus is accessed to fetch the top-ranked suggestion to refine the query. The query interface uses the reformulated query and presents the retrieval set for the reformulated query to the

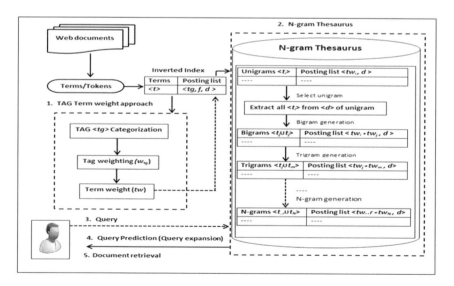

Fig. 3.1 Functional units of N-gram Thesaurus construction

user. The experimental results have shown that the N-gram Thesaurus and weight assigning mechanism are effective and retrieve relevant information with higher precision of the retrieval.

3.4 TAG Term Weight (TTW)

The Internet has large number of web pages, and these web pages are with HTML TAGs. The TAGs of HTML documents reflect the importance, properties and characteristics of text/term within the TAG. These TAG can also be viewed as objects embedded in HTML document and can be modelled as a tree. The extended structure of the tree modelled as Distributed Object Model with a hierarchical structure with clear relationship among each object in a HTML page. As a result, the meaning and semantics of each element of the tree are mapped in syntactical context. A typical HTML document is represented as DOM tree in Fig. 3.2.

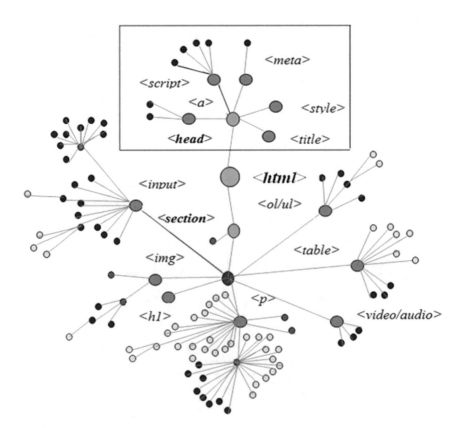

Fig. 3.2 DOM tree for HTML TAGs

The semantic relationship among various TAGs can be effectively captured in HTML version 5, and thus, the DOM tree is presented for HTML version 5. There are one hundred and five (105) TAGs, which are used for building the HTML pages. The root element of the tree is <HTML> and is categorized as <head> and <Section>. The root is an important element, and thus, it is assigned 100% weight. Similarly, the weights for other elements are distributed from the root.

The elements of HTML pages are categorized into metadata content, flow content, phrasing content, embedded content, interactive content, etc. The contents of the element are described as content-element model. The metadata and flow control models are considered as important to effectively define HTML documents. The information of metadata is described in <head> header, and flow content-related information is described in <section> header. The information on text, lists audio/video, images, tables, input controls, style, script are represented by both metadata and flow control HTML TAG and is shown below.

<head>: {*<title>,<a>,<meta>,<style>,<script>*}

<section>: {*<h1>,<p>,<img>,<video>/<audio>,<table>,<input>,<ol>/<ul>*}

It is observed that three-fourth of the TAG elements in the HTML document is present in *<section>* field and one-fourth is from *<head>* section of the HTML. As a result, there are more number of fields in *<section>* header and assigned 25% weights to the TAGS belong to this header. Similarly, *<head>* section is assigned 75% of weights for the TAGs present under this section. These weights are denoted as w_h and w_s and are presented in Eq. (3.1).

$$W = (w_h + w_s) \tag{3.1}$$

The syntactical context of each of the TAGs present in HTML pages is analysed based on the significance. The significance of the TAGs is captured using information present within the TAG structure as presented in (Arnaud and Elena, 2003) of IBM. The metadata content plays important role in setting up the behaviour and relationship among the HTML documents. The information content present within these TAGs in terms of syntactical context is estimated, and the degree of significance is calculated. For example, the title of a HTML pages has syntactical context, and thus, the <title> TAG has more significance. The HTML page is described with author information, data f modification of the pages, etc., and is presented in <metadata> section. In a similar way, the significance is calculated for all the TAGs and it is depicted in Fig. 3.3. It is noticed from the figure that the significant values (k) of a TAG denote the degree of syntactical context of the particular TAG. The number of TAGs available in <head> and <Section> decides the significant graph and relationship. The significant scale of <head> section is 1–5 since there are five TAGs. Similarly, the significant scale of <section> is 1–7 as there are seven TAGs. The significant scale is presented in Fig. 3.3b.

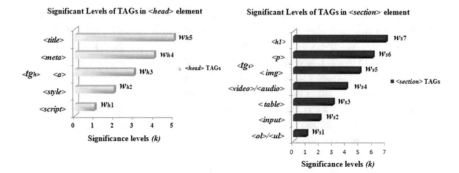

Fig. 3.3 Significant scale **a** *<head>* field and **b** *<section>*

Equations (3.2) and (3.3) assign significant values to the <head> and <section> TAGs. As mentioned earlier, 25% weights are distributed to TAGs under <head> and 75% weights are distributed to TAGs under <section>. In Eqs. (3.2) and (3.3), $w_{hk}(tg_h)$ and $w_{sk}(tg_s)$ are weights assigned for TAGs in *<head>* and *<section>* fields, w_h and w_s are weights of *<head>* and *<section>* fields, and k represent the significant values of various TAGs in *<head>* and *<section>* field.

$$w_{hk}(tg_h) = \left[k / \sum_{k=1}^{5} k \right] \times w_h \tag{3.2}$$

and

$$w_{sk}(tg_s) = \left[k / \sum_{k=1}^{7} k \right] \times w_s \tag{3.3}$$

Assigning the weights to these TAGs from their parent TAGs, i.e. *<head>* and *<section>* fields is represented in Eq. (3.4).

$$\left[w_h = \sum_{k=1}^{5} w_{hk}(tg_h) \right] \text{ and } \left[w_s = \sum_{k=1}^{7} w_{sk}(tg_s) \right] \tag{3.4}$$

- $w_h(<head>) = \{ w_{h5}(<title>) + w_{h4}(<meta>) + w_{h3}(<a>) + w_{h2}(<style>) + w_{h1}(<script>) \}$
- $w_s(<section>) = \{ w_{s7}(<h_1>) + w_{s6}(<p>) + w_{s5}(<img>) + w_{s4}(<video>/<audio>) + w_{s3}(<table>) + w_{s2}(<input>) + w_{s1}(<ol>/<ul>) \}$

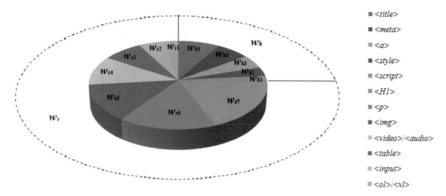

Fig. 3.4 Weight for TAGs under *<head>* and *<section>*

In Fig. 3.4, the pi chart shows the distribution of weights.

Figure 3.5 presents the hierarchical level of all the TAGs in HTML documents. The <head> and <section> are in the second level, and children of each TAG is percolating down with corresponding weight. Finally, the total weight is calculated by summing up weights of TAGs at each level of hierarchical tree.

Equations (3.5) and (3.6) distribute the weights to various TAGs. In these equations, l is leaf node, nl is non-leaf node, k_C is significant value of Children TAG, R is number of children TAG present in the leaf node, and k is significant value of TAGs in *<head>* and *<section>* fields. The total weight is calculated using Eq. (3.7).

$$w_{hkk_c}\left(tg_h^C\right) = \begin{cases} (w_{hk}(tg_h)/R), & \text{if } \left(tg_h^C\right) = l \\ \left[k_C / \sum_{\forall tg_h^C \in tg_h} k_C\right] \times w_{hk}(tg_h), & \text{if } \left(tg_h^C\right) = nl \end{cases} \qquad (3.5)$$

and

$$w_{skk_c}\left(tg_s^C\right) = \begin{cases} (w_{sk}(tg_s)/R), & \text{if } \left(tg_s^C\right) = l \\ \left[k_C / \sum_{\forall tg_s^C \in tg_s} k_C\right] \times w_{sk}(tg_s), & \text{if } \left(tg_s^C\right) = nl \end{cases} \qquad (3.6)$$

$$\text{weight}_{\text{subscripts}} : w_{hkk_C k_{GC} k_{GGC}} \qquad (3.7)$$

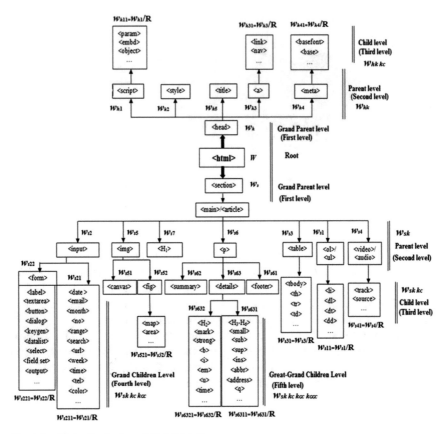

Fig. 3.5 Weight distribution of HTML TAGs

3.5 N-Gram Thesaurus for Query Expansion

As mentioned earlier, information present in the inverted index is effectively used for building the N-gram Thesaurus. The documents crawled from WWW is denoted as D, and T is the term in the document D. The relationship between D and T can be considered as labelling (L) and is shown in Eq. (3.8).

$$L : T \times D \rightarrow \{\text{True}, \text{False}\} \tag{3.8}$$

For each term $<t_i>$, the posting list is extracted from the inverted index having $<d,$ $tw_i>$ information. Here, t_i is a term, d is a document id, and tw_i is the term weight. It is known that terms t_i are physically present in most of the document and is presented as $t_i \in T$ present in a document $d \in D$, if $L : (t_i, d) = \text{True}$. Given an Unigram term $<t_i^U>$, such that $t_i^U \subseteq t_i$ is compared with $\{t_1, t_2, \ldots, t_i\} \in T$ to find whether the term is present in the inverted index. Here, U is Unigram. The $<t_i>$ having matched

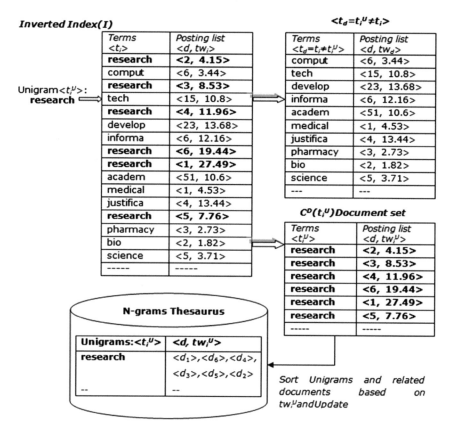

Fig. 3.6 Unigram Thesaurus

with $<t_i^U>$ are considered together with posting list $<d, tw_i>$. These documents are reordered using the term weights $<tw_i>$ as reference, and higher weighted terms top the ranking list. Thus, given an Unigram $<t_i^U>$, $<d, tw_i^U>$ is the posting list. Here, tw_i^U is weight of Unigram term. The ordered documents $<d, tw_i^U>$ are ranked for a given Unigram $<t_i^U>$ and are updated into N-gram Thesaurus $C^D(t_i^U)$ as shown in Eq. (3.9).

$$C^D\left(t_i^U\right) = \left\{\left(t_i^U, d, tw_i^U\right) \Big| t_i^U \in t_i \in T,\ tw_i^U \in TW_i^U, d \in D, \forall t_i^U \subseteq t_i \subseteq T, \left(t_i^U, d\right) = \text{True}\right\}$$

(3.9)

Similarly, all the single terms present in the inverted index are processed with term and corresponding weight. The Unigram Thesaurus is updated from the posting list information. The Thesaurus is updated based on the term weight (TW_i^U). For better understanding, the entire procedure is depicted in Fig. 3.6. In this figure, the term 'research' is considered as Unigram. From the posting list and inverted index,

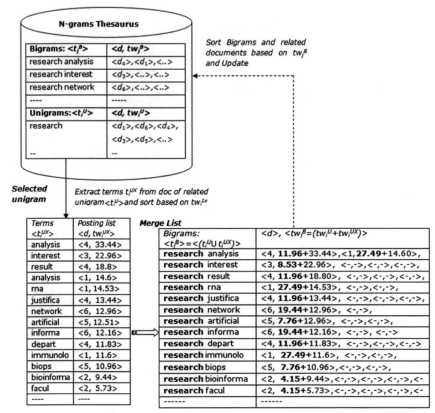

Fig. 3.7 Bigram Thesaurus

only the information related to 'research' is considered as Unigram. The Unigram Thesaurus is sorted based on term weight and used for refining the query.

$$C^D\left(t_j^B\right) = tjB, d, twjBtjB \in tiU \in T \in D, \forall tjB = tiU \cup tiUX, \forall twjB$$
$$= twiU + twiUX, tjB, d = \text{True} \tag{3.10}$$

The procedure of constructing the bigram is shown in Fig. 3.7. The term 'research' is again considered as reference. All the terms having research as co-occurrence term are fetched along with weight, and a merged list is generated. The similar approach is followed to construct N-gram Thesaurus. It is represented in Eq. (3.11) and depicted in Fig. 3.8.

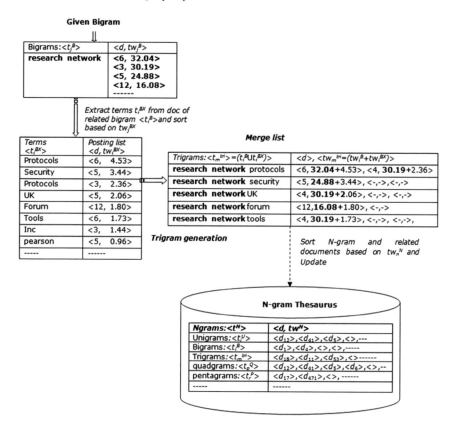

Fig. 3.8 Procedure to generate N-gram Thesaurus

$$
C^D\left(t_n^N\right) = \left(\begin{array}{c} \left\langle t_n^N, d, tw_n^N\right\rangle | t_n^N \in t_{(n-1)}^{(N-1)} \in t_i \in T \in D, \forall t_n^N = t_{(n-1)}^{(N-1)} \cup t_{(n-1)}^{(N-1)X} \\ \text{and } \forall tw_n^N = tw_{(n-1)}^{(N-1)} + tw_{(n-1)}^{(N-1)X}, \left(t_n^N, d\right) = \text{True} \end{array}\right)
$$

$$(3.11)$$

3.6   Algorithm

The N-gram Thesaurus generation algorithm for expanding the query is presented below.

Algorithm:N GRAM THESAURUS GENERATION FOR QUERY EXPANSION

---

1. QUERY-EXPANSION (Inverted-index(I)={<t_i>,<d, tw_i>}, Unigram<t_i^U>)

2. Consider Unigram <t_i^U> and match with {t_i} of I
   (a) Extract matched t_i with posting list <d, tw_i> from I
   (b) Rank documents (d) based on term weights <tw_i>
   (c) Matched terms are considered as Unigrams <t_i^U> along with their posting list <d, tw_i^U>

3. Repeat the process for various terms of I

4. Build Thesaurus by updating Unigrams with posting list ($C^D(t_i^U)$ = <t_i^U><d, tw_i^U>)

5. Rank all Unigrams in Thesaurus based on TW_i^U

6. Bigram expansion: For a given Unigram ($C^D(t_i^U)$ = <t_i^U><d, tw_i^U>)
   (a) Extract all the terms {t_i^{UX}} from the related doc <d> of a given Unigram
   (b) Build Bigrams <t_j^B>=<$t_i^U \cup \{t_i^{UX}\}$> and compute Bigram weights <tw_j^B>=<$tw_i^U + tw_i^{UX}$>
   (c) Rank documents <d> based on Bigram term weights {tw_j^B}

7. Repeat the process for various Unigrams of Thesaurus

8. Update Thesaurus with Bigrams along with their posting list ($C^D(t_j^B)$=<t_j^B><d, tw_j^B>)

9. Rank all Bigrams based on TW_j^B

10. (N+1) gram expansion: repeat steps 6 to 9 using N grams

11. Update (N+1) grams into Thesaurus for expansion

```
12. If (user gives N gram query) then
        Display list of Predicted (N+1) grams as
        suggestion from Thesaurus

            If (User selects suggested (N+1) grams and
            expands query) then
                Display ranked documents for an expanded
                Query

        Else
            Display ranked documents for an original
            user query(unexpanded)
```

3.7 Experimental Results

The N-gram Thesaurus approach is evaluated on three benchmark datasets such as GOV.txt, clueweb09.txt and WT10g.txt. These datasets are available at http://dl.d ropbox.com/u/84084/GOV2.txt, http://dl.dropbox.com/u/84084/clueweb09.txt and http://dl.dropbox.com/u/84084/WT10g.txt. The content of datasets in terms of number of features, number of documents and average query length is presented in Table 3.1.

3.7.1 Performance Evaluation of Query Reformulation and Expansion

In this section, the experimental results are presented. The result is compared with BoCo, KLDCo (Perez-Aguera and Lourdes-Araujo 2008) Co approach (Van Rijsbergen 1977), Bose–Einstein statistics weighting model (Bo1) (Macdonald et al. 2005) and KLD (Cover and Thomas 1991). The user query and expanded terms are presented in Table 3.2 for all the methods.

Table 3.1 Information on benchmark dataset

Features	Dataset used		ClueWeb09B
	WT10g	GOV2	
Total # of queries	100	149	98
# of docs	1,692,096	25,205,179	50,220,423
Avg. query length	2.31	2.59	1.96

Table 3.2 Clueweb09B: expanded queries

Approaches	Queries and expanded terms
	Cloning
KLDCo	Cloning humans, cloning animals, cloning dinosaurs, cloning insects, cloning methods, cloning software
BoCo	Cloning ford, cloning marijuana, cloning humans, cloning dinosaurs
TTW	Cloning humans, cloning animals, cloning plants, cloning dinosaurs, cloning insects, cloning asexually, cloning DNA
User selection	Cloning humans, cloning asexually, cloning DNA
	Diet
KLDCo	Diet pill, diet plan, diet coke, diet recipes, diet obese, diet problems, diet weight loss
BoCo	Diet meals, diet coke, diet plan, diet pills, dietician
TTW	Diet practice, diet meals, diet plan, diet pills, dietician, diet types, diet coke
User selection	Diet practice, diet types, diet pill, diet plan, diet meals
	Mobile phone
KLDCo	Mobile phone services, mobile phone seniors, mobile phone reviews, mobile phone mechanism
BoCo	Mobile phone deals, mobile phone companies, mobile phone tracking, mobile phone reviews
TTW	Mobile phone deals, mobile phone banking, mobile phone network, mobile phone accessories, mobile phone reviews, mobile phone companies
User selection	Mobile phone deals, mobile phone reviews, mobile phone accessories

Table 3.3 Query expansion and user analysis

Analysis features	KLDCo	BoCo	TTW
# of suggestions provided in each approach (average)	4.4	3.3	6.0
# of suggestions selected by the user in each approach (average)	1.47	1.2	2.7
% of suggestions selected in each approach	33%	36%	45%
% of all selections from each approach	27%	22%	50%

The experiment is also conducted for users query. For each query, the choice of the user on expanded query for various dataset is presented in Table 3.3.

The users are given the expanded query terms for retrieving relevant documents. For all the expanded queries, the retrieval set is more relevant.

Table 3.4 Baseline performance on Clueweb09B, WT10g and GOV2 datasets

Dataset	P@5	P@10	MAP	MRR
Clueweb09B	0.323	0.305	0.282	0.336
WT10g	0.362	0.318	0.221	0.412
GOV2	0.331	0.307	0.208	0.487
Average score	0.338	0.310	0.237	0.411

Table 3.5 Comparative performance of TTW

Dataset	P@5	P@10	MAP	MRR
Clueweb09B	0.442 (+0.119)	0.412 (+0.107)	0.361 (+0.079)	0.531 (+0.195)
WT10g	0.656 (+0.294)	0.527 (+0.209)	0.404 (+0.183)	0.695 (+0.283)
GOV2	0.625 (+0.294)	0.618 (+0.311)	0.502 (+0.294)	0.738 (+0.251)
Average score	0.574 (+0.102)	0.519 (+0.128)	0.422 (+0.118)	0.654 (+0.243)

3.7.2 Performance of TTW Approach

For the evaluation purpose, the original query term is considered along with frequency of occurrence. This is treated as baseline as the frequency of occurrence is descriptive about the term capacity and complete metric. The logical AND combination of query terms is used as query. While analysing the retrieved documents, all the terms appearing in query appear anywhere in the document. The precision of retrieval, MAP and MRR are used as performance metric, and the result is presented in Table 3.4 for various benchmark datasets.

Similarly, Table 3.5 presents the improvement in performance by using TTW along with N-gram Thesaurus. Further, Table 3.6 presents the comparative performance of TTW with N-gram Thesaurus.

The average improvement gain against baseline average scores for combined dataset is presented in Table 3.7. The difference gain is presented in Table 3.8.

Based on all the experiments presented in this section, it is observed that the TTW along with N-gram Thesaurus expands the query term. While these query terms are presented as query in the retrieval interface, the retrieval set contains the most relevant documents. The comparative results also show that the TTW has outperformed all the methods.

Table 3.6 Performance comparison of various approaches for various datasets against baselines

Approach	Clueweb09B			WT10g			GOV2		
	P@10	MAP	MRR	P@10	MAP	MRR	P@10	MAP	MRR
Baseline	**0.305**	**0.282**	**0.336**	**0.318**	**0.221**	**0.412**	**0.307**	**0.208**	**0.487**
TTW	0.412 (+0.107)	0.361 (+0.079)	0.531 (+0.195)	0.527 (+0.209)	0.404 (+0.183)	0.695 (+0.283)	0.618 (+0.311)	0.502 (+0.294)	0.738 (+0.251)
Co	0.310 (+0.005)	0.290 (+0.008)	0.375 (+0.039)	0.356 (+0.038)	0.323 (+0.102)	0.426 (+0.014)	0.316 (+0.009)	0.302 (+0.094)	0.465 (+0.022)
KLD	0.318 (+0.013)	0.314 (+0.031)	0.401 (+0.065)	0.328 (+0.010)	0.319 (+0.098)	0.430 (+0.018)	0.320 (+0.013)	0.314 (+0.106)	0.522 (+0.035)
BoCo	0.312 (+0.007)	0.310 (+0.028)	0.444 (+0.108)	0.569 (+0.251)	0.483 (+0.262)	0.709 (+0.297)	0.445 (+0.138)	0.434 (+0.226)	0.683 (+0.196)
KLDCo	0.319 (+0.013)	0.315 (+0.032)	0.489 (+0.153)	0.479 (+0.161)	0.401 (+0.180)	0.503 (+0.091)	0.413 (+0.106)	0.459 (+0.251)	0.622 (+0.135)

Table 3.7 Gain improvement

Combined dataset (average Scores)	P@5	P@10	MAP	MRR
Baseline	**0.338**	**0.310**	**0.237**	**0.411**
TTW	0.574 (+0.236)	0.519 (+0.209)	0.422 (+0.185)	0.654 (+0.243)
Co	0.356 (+0.018)	0.328 (+0.018)	0.305 (+0.068)	0.422 (+0.011)
KLD	0.374 (+0.036)	0.322 (+0.002)	0.315 (+0.076)	0.451 (+0.040)
BoCo	0.502 (+0.164)	0.442 (+0.132)	0.409 (+0.172)	0.612 (+0.201)
KLDCo	0.410 (+0.072)	0.401 (+0.091)	0.391 (+0.154)	0.538 (+0.127)

Table 3.8 Difference in the improvement gain of comparative approaches with TTW approach

Combined dataset (average scores)	P@5	P@10	MAP	MRR
TTW	0.574	0.519	0.422	0.654
BoCo	0.502 (−0.072)	0.442 (−0.077)	0.409 (−0.013)	0.612 (−0.042)
KLDCo	0.410 (−0.164)	0.401 (−0.118)	0.391 (−0.031)	0.538 (−0.116)

3.8 Conclusion

The query refinement and reformulation are useful techniques to improve the searching experience. The characteristics of HTML TAGs and its importance are estimated using the syntactical context. Based on the importance, suitable weights are derived and assigned to the terms appearing between the TAGs. N-gram Thesaurus is constructed using the information from inverted index and posting list. The weights of the terms are used for refining the query and ranking the query suggestion. The performance is evaluated on well-known benchmark datasets. Overall, the TTW along with N-gram Thesaurus has outperformed all the methods.

References

Arnaud, L. H., & Elena, L. (2003) (IBM) Discover key features of DOM level 3 core, part 1, manipulating and comparing nodes, handling text and user data.

Blei, D. M., Ng, A. Y., & Jordan, M. I. (2003). Latent Dirichlet allocation. *Journal of Machine Learning Research, 3*, 993.

Cover, T. M., & Thomas, J. A. (1991). *Elements of information theory*. New York, NY, USA: Wiley-Interscience.

Cucerzan, S., & Brill, E. (2004). Spelling correction as an iterative process that exploits the collective knowledge of web users. In *Proceedings of EMNLP* (p. 293).

Francesco, C., Massimo, D. S., Luca, G., & Paolo, N. (2013). A query expansion method based on a weighted word pairs approach. In *Proceedings of 4th IIR Workshop 2013*. Pisa, Italy: National Council of Research Campus.

Gong, Z., Cheang, C., & Hou, U. L. (2005). Web query expansion by wordnet. In *Proceedings of Database and Expert Systems Applications* (pp. 166–175). Berlin/Heidelberg: LNCS, Springer.

Griffiths, T. L., Steyvers, M., & Tenenbaum, J. B. (2007). Topics in semantic representation. *Psychological Review, 114*(2), 211.

Jing, Y., & Croft, W. B. (1994). An association Thesaurus for information retrieval. In *Proceedings of RIAO 94 Conference* (pp. 146–160).

Kaptein, R., & Kamps, J. (2009). *Advances in focused retrieval. Chapter finding entities in Wikipedia using links and categories* (pp. 273–279). Berlin, Heidelberg: Springer-Verlag.

Kilgarriff, A. (2007). Googlelology is bad science. *Journal of Computational Linguistics, 33*(1), 147.

Li, Y., Luk, W. P. R., Ho, K. S. E., & Chung, F. L. K. (2007). Improving weak ad-hoc queries using Wikipedia as external corpus. In *Proceedings of 30th ACM SIGIR Conference on Research and Development in Information Retrieval SIGIR '07* (pp. 797–798). New York, USA.

Lin, H. C., Wang, L. H., & Chen, S. M. (2005). A new query expansion method for document retrieval by mining additional query terms. In *Proceedings of International Conference on Business and Information*. Hong Kong, China.

Macdonald, C., He, B., Plachouras, V., & Ounis, I. (2005). University of Glasgow at TREC 2005: Experiments in terabyte and enterprise tracks with terrier. In *Proceedings of 14th Text REtrieval Conference (TREC 2005)*.

Manning, C. D., Raghavan, P., & Schutze, H. (2008). *Introduction to information retrieval*. Cambridge University Press.

Marianne, H., Nadja, N., & Carolin, B. (Eds.). (2007). *Corpus linguistics and the web: In literary and linguistic computing* (p. 305). Amsterdam/New York: Radopi.

Martin-Bautista, M. J., Sanches, D., Chamorro-Martinez, J., Serrano, J. M., & Vila, M. A. (2004). Mining web documents to find additional query terms using fuzzy association rules. *Fuzzy Sets and Systems, 148*(1), 85.

Metzler, D., & Croft, W. B. (2007). Latent concept expansion using markov random fields. In *Proceedings of 30th Annual International ACM SIGIR Conference on Research and Development in Information Retrieval* (p. 311).

Milne, D. N., Witten, I. H., & Nichols, D. M. (2007). A knowledge based search engine powered by Wikipedia. In *Proceedings of 16th ACM Conference on Information and Knowledge Management. CIKM '07* (pp. 445–454). New York, USA.

Perez-Aguera, J. R., & Lourdes-Araujo. (2008). Comparing and combining methods for automatic query expansion. *Advances in Natural Language Processing Research in Computing Science, 33*, 177–188.

Qiu, Y., & Frei, H.-P. (1993). Concept based query expansion. In *Proceedings of 16th ACM SIGIR Conference on Research and Development in Information Retrieval* (pp. 160–169). New York, NY, USA.

Smeaton, A. F., Kelledy, F., & O'Donnell, R. (1995). Thresholding posting lists, query expansions with wordnet and pos tagging of Spanish. In *Proceedings of 4th Text REtrieval Conference (TREC-4)* (pp. 373–390).

Van Rijsbergen, C. J. (1977). A theoretical basis for the use of cooccurrence data in information retrieval. *Journal of Documentation, 33,* 106–119.

Voorhees, E. M. (1994). Query expansion using lexical-semantic relations. In *Proceedings of 17th ACM SIGIR Conference on Research and Development in Information Retrieval* (pp. 61–69). New York, USA: Springer-Verlag Inc.

Wang, C., Yajun, D., Zhang, P., & Han, B. (2010). A term-reweighting method for query expansion. *Journal of Computational Information Systems, 6*(11), 3779.

Xu, Y., Jones, G. J., & Wang, B. (2009). Query dependent pseudo-relevance feedback based on Wikipedia. In *Proceedings of 32nd InterNational ACM SIGIR Conference on Research and Development in Information Retrieval SIGIR '09* (pp. 59–66). New York, USA.

Chapter 4
Smooth Weighted Colour Histogram Using Human Visual Perception for Content-Based Image Retrieval Applications

4.1 Introduction

Various techniques have been proposed in academia and industry for image retrieval applications. These approaches can roughly be classified into three categories such as text-based retrieval, content-based retrieval and semantic-based retrieval. In the text-based retrieval technique, each image has a number of keywords for describing the image and the keyword-based matching is performed to retrieve relevant images. In content-based image retrieval applications, various well-known low-level features like colour, texture and shape are extracted for describing the image semantics. In the semantic-based retrieval technique, semantics are used to retrieve the relevant images. In recent years, attention is focused by researchers on content-based image retrieval (CBIR), which is a sub-problem of content-based retrieval. It is noticed that the size of the image database used in image retrieval is increasing exponentially, and hence, it is necessary to propose and use effective tools for retrieving relevant images. Well-known and most popular image retrieval systems are QBIC (Niblack et al. 1993), NeTra (Ma and Manjunath 1997), PicToSeek (Gevers and Smeulders 2000), Blobworld (Carson et al. 1999), etc. Human visual system plays a vital role in the colour theory of image retrieval applications as human eyes first captures colour. There are two types of cells in human retina, namely the rod cells and the cone cells, and are responsible for the sensitivity of the light. Rod cells are responsible for grey vision, and cone cells are responsible for colour vision. Colour vision results from the action of three cone cells with different spectral sensitivities at red (R), green (G) and blue (B) of the visible light spectrum. The peak sensitivities of these three cone cells are located approximately at 610, 560 and 430 nm, respectively. When particular wavelength of a light is incident on the eye, the cells are stimulated to different degrees, and the ratio of the activity in the three cells results in the perception of a particular colour. Each colour is therefore coded in the nervous system by its own ratio of activity in the three types of cone cells. For a given wavelength of light,

© Springer Nature Singapore Pte Ltd. 2018
S. G. Shaila and A. Vadivel, *Textual and Visual Information Retrieval using Query Refinement and Pattern Analysis*, https://doi.org/10.1007/978-981-13-2559-5_4

human colour perception is determined by which combination of cones is excited and by how much (Vadivel et al. 2008).

In this chapter, the goal is to retrieve set of images that are similar to a given query image. In image retrieval applications, it has been observed that the colour histogram-based approach is well suited, since colour matching generates the strongest perception of similarity to the human eye. It is often represented in the form of a histogram, which is a first-order statistical measure that captures the global distribution of colour in a given image. This can be represented in various forms such as colour histogram (Swain and Ballard 1991), colour moments and cumulative colour histogram (Stricker and Orengo 1995). Cumulative colour histogram uses the spatial relationship between histogram bins. A colour histogram may be generated in the RGB colour space, HSV colour space, YCbCr colour space, etc. Since the RGB colour space is having some drawbacks such as it does not explicitly distinguish between colour and intensity components, other colour spaces like HSV, YCbCr, which separate saturation and intensity components, are frequently used. RGB values by suitable conversions may be computed to get the values for each component in HSV colour space (Vadivel et al. 2003). It is also possible to generate three separate histograms, one for each channel, and concatenate them into one (Jain and Vailaya 1996). Some of the recently proposed colour representatives are colour saliency histogram (Gong et al. 1998), which supports human visual attention principle, and based on the features of colour, orientation and intensity and followed by the difference of Gaussians and normalization processing, the comprehensive saliency map of an image is generated. Recently, circular ring histogram (Wang 2009) has been proposed, which has the spatial information. The image is segmented initially using group of circular rings, and then, the histogram is constructed using the segmented rings. The statistical feature of image blocks has been extracted for representing the colour of blocks, which is named as order-based block colour feature (Wang and Qin 2009); in this approach, image is divided into 48 blocks and the feature is extracted. Since histogram components store the number of pixels having similar colours, it may be considered to be a signature of the complete image represented by a feature vector. In this chapter, a new colour histogram construction scheme based on human colour perception is presented and NBS distance is used to capture the colour information in high-dense background of images for retrieval applications. Pixel weight is distributed to the neighbouring bins based on NBS distance, and the performance of the MHCPH approach is encouraging while retrieving high-dense background images. During retrieval, histogram of query image is compared with histograms in feature database. The histogram is a point in the n-dimensional vector space, and a distance measure like the Manhattan distance is used for the comparison and ordering of these vectors.

The outline of this chapter is as follows. Section 4.2 presents the related works, and modified human colour perception histogram (MHCPH) technique is explained in Sect. 4.3. In Sect. 4.4, the experimental results are presented and concluded the chapter in the last section.

4.2 Reviews on Colour-Based Image Retrieval

In colour-based image retrieval, there are primarily two methods: one based on colour layout (Smith and Chang 1996) and the other based on colour histogram (Deb and Zhang 2004). In the colour layout approach, two images are matched by their exact colour distribution. This means that two images are considered close if they not only have similar colour content, but also have similar colour in approximately the same positions. In the second approach, each image is represented by its colour histogram. A histogram is a vector, whose components represent a count of the number of pixels having similar colours in the image. Thus, a colour represents to be a signature extracted from a complete image. Colour histograms extracted from different images are indexed and stored in a database. During retrieval, the histogram of a query image is compared with the histogram of each database image using a standard distance metric like Euclidean distance or Manhattan distance (Vadivel et al. 2003). Since colour histogram is a global feature of an image, the approaches based on colour histogram are invariant to translation, rotation and scale.

Various techniques have been proposed to represent the colour of an image (Deng et al. 2001; Han and Ma 2002; Kender 1976; Lei et al. 1999; Nezamabadi-pour and Kabir 2004; Shih and Chen 2002). Gevers and Stokman (2004) have proposed a histogram, and it is used in object recognition problem. A variable kernel density estimation is used in this approach to construct colour invariant histograms. Uncertainty associated with colour variant values of hue and normalized RGB colour is computed and is used to determine the kernel sizes. In pyramid histogram of topics (PHOTO) (Lu et al. 2009), the image is represented to suit the classification requirement and image is partitioned into hierarchical cells. The topic histogram is learned using pLSA with EM algorithm. The topic histograms are concatenated over the cells at all levels to form a long vector, i.e. pyramid histogram of topics. In fuzzy, colour histogram (Han and Ma 2002) is constructed by considering the colour similarity of each pixel's colour associated with all the histogram bins through fuzzy-set membership function. A fast approach for computing the membership values based on fuzzy-means algorithm is introduced to identify objects based on the colour histograms. Swain and Ballard (1991) propose a histogram intersection method, which is able to eliminate the influence of colour contributed from the background pixels during the matching process in most cases. Although their method is robust to object occlusion and image resolution, it is still sensitive to illumination changes. It is found that robust histogram construction scheme using the HSV colour space in which a perceptually smooth transition is captured based on the human visual perception of colour (HCPH) can represent semantic information up to a certain degree, due to the complex background (Vadivel et al. 2008). In the DCT histogram quantization (Mohamed et al. 2009), the method extracts and constructs a feature vector of histogram quantization from partial DCT coefficient in order to count the number of coefficients that have the same DCT coefficient over all image blocks. The database image and query image are equally divided into a non-overlapping 8×8 block pixel, each of which is associated with a feature vector of histogram quantization derived directly from discrete cosine transform DCT which results in good results for clear background. In colour saliency histogram (Lei et al. 2009), extraction of salient regions is based on the bottom-up visual attention model. Although it has introduced prior knowledge in the model, the bottom-up attention is only suitable for the primary stage of visual perception, and it has limitations. Order-based block colour feature (Wang and Qin 2009) is one type of image's colour feature. It has an advantage of the local colour statistical information, but it has some bionic traits that this colour feature cannot alone suffice for CBIR. It has to be combined with other features such as shape, texture.

Based on the above discussion, it is noticed that each method used various techniques and none of them uses the human visual perception. However, though HCPH uses human visual perception, it fails to discriminate the background colour with the foreground colour for better retrieval. Thus, in the presented approach, both human visual perception and NBS distance are considered for iteratively distributing the colours to the adjacent bins. It is noticed that this procedure of updating histogram bins effectively discriminates the foreground with the background.

4.3   Human Visual Perception Relation with HSV Colour Space

The HSV colour model is one of the colour models that separate out the luminance component, intensity of a pixel from its chrominance components hue and saturation. This representation is more similar to the human perception of colour through rod and cone cells. Hue represents pure colours and is perceived by the excitation of cone cells when incident light is of sufficient illumination as well as contains a single wavelength. Saturation gives a measure of the degree by which a pure colour is diluted by white light. Light containing such multiple wavelengths causes different excitation levels of the cone cells resulting in a loss of colour information. For light with low illumination, corresponding intensity value in the HSV colour space is also low. Only the rod cells contribute to visual perception at low-intensity illumination with little contribution from the cone cells. Pixels in a image can be represented with the combination of hue, saturation and intensity value in the HSV colour space. It is a three-dimensional hexacone representation, and the central vertical axis is intensity, I. Hue is an angle in the range $[0, 2\pi]$ and is relative to the red axis with red at angle 0, green at $2\pi/3$, blue at $4\pi/3$ and red again at 2π (Gong et al. 1998), respectively. Saturation, S, is measured as a radial distance from the central axis with a value between 0 at the centre to 1 at the outer surface (Vadivel et al. 2008). This is represented in Fig. 4.1.

The pixel colour is approximated by its intensity or by its hue, say while the intensity is low and saturation is high, a pixel colour is very much close to the grey colour. Similarly, for other combinations of intensity and saturation, the pixel value can be approximated the other ways. Therefore, the saturation and intensity values of

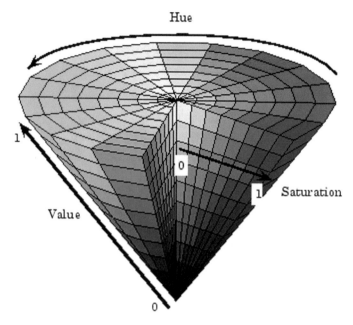

Fig. 4.1 HSV colour model

a pixel are used to determine whether it can be treated as a true colour pixel or a grey colour pixel. It is emphasized that this approach treats the pixels as a distribution of 'colours' in an image where a pixel may be of a 'gray colour' (i.e. somewhere between black and white, both inclusive) or of a 'true colour' (i.e. somewhere in the red, green, blue, red spectrum). The reason is that for human an image is a collection of points having colours—red, yellow, green, blue, black, gray, white, etc.

4.4 Distribution of Colour Information

In the MHCPH work, colour information of each pixel is converted into HSV colour space. An image pixel contains true colour components, in which the dominant factor is hue. Intensity is the dominant factor for grey colour components. In the first step, the pixel colour information is separated into true colour components and grey colour components using a weight function (Vadivel et al. 2008) given in Eqs. (4.1) and (4.2)

$$W_H(S, I) = \begin{cases} S^{r\,1(255/I)^{r2}} & \text{for } I \neq 0 \\ 0 & \text{for } I = 0 \end{cases} \qquad (4.1)$$

$$W_I(S, I) = 1 - W_H(S, I) \qquad (4.2)$$

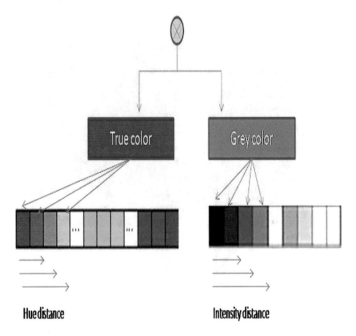

Fig. 4.2 Distribution of true colour and grey colour components

As noticed from Fig. 4.2, the assumption is the true colour component is updated in the first bin of the true colour part and the grey colour component is updated in the first bin of the grey colour part.

4.5   Weight Distribution Based on NBS Distance

The distance between the reference bin and its smoothly distributed corresponding adjacent bins is computed by applying NBS distance formula (Gong et al. 1998) represented by Eq. (4.3). For true colour, ΔS and ΔI remain zero. This is due to the fact that the saturation and intensity distribution difference is invariable or minimal with respect to a single pixel. For grey colour, ΔS and ΔH remain zero, since saturation and hue difference appears to be very minimal or invariable.

$$d(\overrightarrow{x}, \overrightarrow{y}) = 1.2 * \sqrt{2x_2 y_2 \left(1 - \cos\left(\frac{2\pi \Delta H}{100}\right)\right) + \Delta S^2 + (4\Delta I)^2} \qquad (4.3)$$

Table 4.1 represents the NBS distance table which represents human perceptions for various NBS distance values.

Table 4.1 NBS distance table

NBS distance value	Human perception
0–1.5	Almost the same
1.5–3.0	Slightly different
3.0–6.0	Remarkably different
6.0–12.0	Very different
12.0	Different colour

Based on these ranges, weights are assigned to true colour components and grey colour components. It is observed that the immediate adjacent bin close to the reference bin will lie in the distance ranging between 0 and 1.5, which represents the colour difference as 'almost the same' as per the NBS table. Thus, 100% of true colour weight held by reference bin will be distributed to the immediate adjacent bin. As the distance of the adjacent bins increases, the colour difference will also increase proportionately. If the distance lies in the range of 1.6–3.0, it is observed that there is a 'slight colour difference' and is divided it into 3 groups with 15 iterations. Now, the weight is distributed in the range between 99 and 85% of reference true colour weight. When the distance is measured with respect to the adjacent bins in the range of 3.1–6.0, there is a 'remarkable colour difference' range. Hence, the histogram is divided into 6 groups with 30 iterations and weight distribution starts in the range of 84–55% of reference true colour weight. As distance increases, the colour difference will be more, which may lie in the range of 6.1–12.0 and is represented as 'very different colour'. Thus, it is divided into 12 groups with 60 iterations and weight distribution starts in the range of 54–0%; if distance is greater than 12.0, then it represents totally different colour, and there will be no weight distribution or 0% weight distribution, and it is represented in the form of tree in Fig. 4.3.

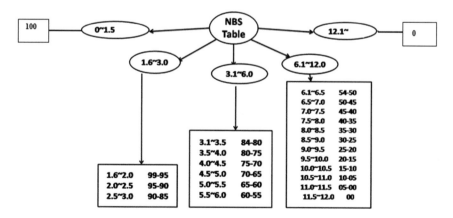

Fig. 4.3 Construction of smooth weight distribution tree

The weight distribution percentage for the true colour to the adjacent bins is calculated using Eq. (4.4)

$$SW_H(S, I) = 100 - [(NBS - 1.5) * 10] \qquad (4.4)$$

During grey colour weight distribution, intensity is distributed smoothly. NBS distance is not applicable for grey colour weight distribution, since the colour difference is very minimum and is not rectified. But distribution of grey colour weight is done by using Eq. (4.5).

$$SW_I(S, I) = 1 - SW_H(S, I) \qquad (4.5)$$

where

NBS	True colour distance
$SW_H(S, I)$	Smooth true colour weights
$SW_I(S, I)$	Smooth grey colour weights

4.6 Algorithm

For each pixel in the image,

Read RGB value

1. Convert RGB - HSV colour space

2. Determine the weight function for true
   colour $W_H(S.I)$ and grey colour $W_I(S.I)$ using
   Eq.(4.1) and Eq.(4.2)

3. Smoothly distribute true colours and grey
   colours among adjacent Bins iteratively.

4. Determine the index of the bins for smoothly
   distributed true colour(Hue) and grey col
   our(Intensity) using MULT_FCTR and DIV_FCTR

5. Compute NBS distance between reference
   colour bin and its corresponding smoothly
   distributed adjacent bins using NBS distance
   formula

6. Based on the computed NBS distance find the
   colour difference Using NBS distance Table.

7. Determine the weight function for smoothly
   distributed true Colour $SW_H(S.I)$ and weight
   function for smoothly distributed grey Colour
   $SW_I(S.I)$ among the bins using Smooth Weight
   Distribution Tree

8. Update the histogram for true colour and grey
   colour as follows:
   a. Smooth True Colour Hist[Round(Smoothly distrib
      uted Hue*MULT_FCTR)]=Smooth True Colour
      Hist[Round(Smoothly distributed
      Hue*MULT_FCTR))]+$SW_H(S.I)$
   b. Smooth grey Colour Hist [Round(2π.MULT_FCTR)]+1
      +Round(Smoothly distributed Intensity/DIV_FCTR)
      =Hist[Round(2π.MULT_FCTR)]+1+Round(Smoothly
      distributed Intensity/DIV_FCTR)+$SW_I(S.I)$

9. Compute Manhattan distance between query image
   and database images to rank the Similarity

Similarly, the distance between the first grey and the adjacent bins is also calculated. Using the distance value, the proportion by which the current pixel colour
overlaps with neighbouring colour is estimated and accordingly the weight of the
bins is updated.

The histogram representation of HCPH is shown in Fig. 4.4a and c. Here, pixel
containing true colour information and grey colour information is grouped into two
separate bins. The Graphical representation of smooth distribution of true colour and
grey colour using MHCPH approach is shown in Fig. 4.4b and d which represents
soft nature of distribution. From the figure, it can be observed that the MHCPH
approach has flat bit representation and thus has more information about the colour
and intensity of image.

4.7   Experimental Results

The effectiveness of the MHCPH is tested in an image retrieval system. The performance evaluation measures for determining the accuracy of such systems include
recall and precision. It is difficult to have a comparative analysis of CBIR schemes
without the benchmark image databases. Without the ground truth, it is hard to calculate recall measures on such databases. In the experiments, coral benchmark database
of about 10,000 images of various classes such as people, vehicle, building, flower is
considered. A set of query images having dense and complex background have been
selected from these 10,000 images for computing recall and precision. Precision is
the ratio of the number of the relevant images retrieved to the total number of images
retrieved.

$$\text{Precision (P)} = \frac{R_r}{T} \qquad (4.6)$$

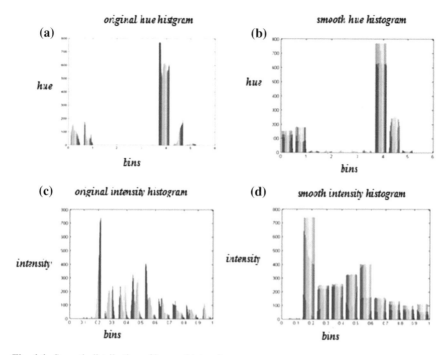

Fig. 4.4 Smooth distribution of hue and intensity

where

R_r is number of relevant images retrieved
T is the total images retrieved

Recall is defined as the ratio of the number of relevant images retrieved to the total number of the relevant images in the database. Recall is a measure of completeness.

$$\text{Recall (R)} = \frac{R_r}{T_r} \tag{4.7}$$

where

R_r is number of relevant images retrieved
T_r is the total number of relevant images in the database

Here, the comparison is done between HCPH and MHCPH method, and the results are shown. top 50 image retrieval has been considered, and the average precision and recall graph for ten different classes are shown in Figs. 4.5 and 4.6. The results for dense background images are obtained progressively.

Average precision versus recall is calculated and is shown in Fig. 4.7. Results are obtained for different values of recall from 0.1 to 1.0 in steps of 0.1, and the corresponding precision was calculated. For all the methods, Manhattan distance

Fig. 4.5 Average precision

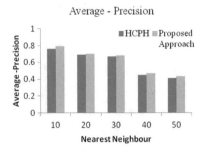

Fig. 4.6 Average recall

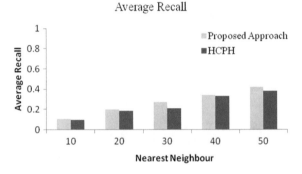

Fig. 4.7 Average precision
versus recall

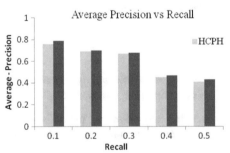

was used as the distance metric. The higher the distance, the lower is the similarity
between a query image and a target image. From the figure, it is observed that for lower
values of recall, the precision is getting higher, which is reaching 80%. Similarly,
for higher value of recall, the precision is comparable and the performance of the
MHCPH method is encouraging.

Further, F measure is used which considers both recall and precision as given in
Eq. (4.8). This is a harmonic mean of recall and precision and is shown in Fig. 4.8.

$$
F = \left(\frac{2PR}{P + R} \right) = \left(\frac{2}{\frac{1}{R} + \frac{1}{P}} \right)
\tag{4.8}
$$

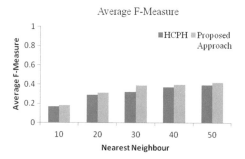

Fig. 4.8 Average *F* measure

Fig. 4.9 Sample retrieval set using MHCPH

Since weight is smoothly distributed to neighbouring bins based on the true colour and gray colour weight, the entire information of the background as well as the object is captured, and thus, the precision retrieval is 80%. Figures 4.9 and 4.10 depict sample retrieval result for both the presented and HCPH histograms. The image on the top centre is the query image, and the rest of the image below is the retrieved image set. In the MHCPH method, background and foreground colours are similar for query image in the top 4 retrieval set. But in HCPH, this appears only in top 2 retrieval set.

HCPH algorithm is compared with two methods proposed in Gevers and Stokman (2004) and colour-based co-occurrence matrix scheme (MCCM) proposed in Shim and Choi (2003). It is observed that the HCPH method outperforms by more than 15%. When presented approach is compared with HCPH, it is found that the presented approach algorithm outperforms between 3 and 5% for dense background images. It is noted that when tested with a large image database having wide variety of

Query Image

Fig. 4.10 Sample retrieval set using HCPH

images, precision of all the image retrieval techniques deteriorates considerably. This is balanced in the results when tested on 10,000 images, which have been categorized into different groups. A high percentage of these relevant images are retrieved for each query, resulting in high precision and recall. However, even with uncontrolled image database, the performance of the presented method is comparatively better than the other approaches.

4.8   Conclusion

In this chapter, a new colour histogram construction technique is presented. The colour features of each pixel of the image are smoothly distributed to the neighbouring bins iteratively. This representation of colour distribution of an image captures the foreground and background information of an image effectively. This is due to the fact that this approach uses both human visual perception and NBS distance measure. The performance of the presented approach is encouraging in benchmark datasets, and it is found that the precision of retrieval is higher for background complex images. For future work, the rate of retrieval can be improved by adding more iterations and adding extra features such as texture and shape with the colour to improve the relevance by effectively balancing both of these parameters.

References

Carson, C., Thomas, M., Belongie, S., Hellerstein, J. M., & Malik, J. (1999). Blobworld: A system for region-based image indexing and retrieval. In *Proceedings of Third International Conference on Visual Information Systems* (pp. 217–225).

Deb, S., & Zhang, Y. (2004). An overview of content-based image retrieval techniques. In *Proceedings of 18th International Conference on Advanced Information Networking and Applications* (Vol. 1, pp. 59–64).

Deng, Y., Manjunath, B. S., Kenney, C., Moore, M. S., & Shin, H. (2001). An efficient colour representation for image retrieval. *IEEE Transactions on Image Processing, 10,* 140–147.

Gevers, T., & Smeulders, A. W. M. (2000). PicToSeek: Combining colour and shape invariant features for image retrieval. *IEEE Transactions on Image Processing, 9,* 102–119.

Gevers, T., & Stokman, H. M. G. (2004). Robust histogram construction from colour invariants for object recognition. In *IEEE Transactions on Pattern Analysis and Machine Intelligence* (Vol. 26, pp. 113–118).

Gong, Y., Proietti, G., & Faloutsos, C. (1998). Image indexing and retrieval based on human perceptual colour clustering. In *Proceedings of IEEE Computer Society Conference on Computer Vision and Pattern Recognition* (pp. 578–583).

Han, J., & Ma, K.-K. (2002). Fuzzy colour histogram and its use in colour image retrieval. *IEEE Transactions on Image Processing, 2,* 944–952.

Jain, A., & Vailaya, A. (1996). Image retrieval using colour and shape. *Pattern Recognition, 29,* 1233–1244.

Kender, J. R. (1976). *Saturation, hue and normalised colour: Calculation, digitisation and use* (Computer Science Technical Report). Pittsburg, USA: Carnegie-Mellon University.

Lei, Z., Fuzong, L., & Bo, Z. (1999). A CBIR method based colour-spatial feature. In *Proceedings of IEEE Region 10 Annual International Conference on TENCON 99* (pp. 166–169). Cheju Island, South Korea.

Lei, Y., Shi, Z., Jiang, X., Li, Q., & Chen, D. (2009). Image retrieval based on colour saliency histogram. In *International Symposium on Computer Network and Multimedia Technology* (pp. 1–4).

Lu, F., Yang, X., Zhang, R., & Yu, S. (2009). Image classification based on pyramid histogram of topics. In *Proceedings of IEEE International Conference on Multimedia and Expo, ICME 2009* (pp. 398–401).

Ma, W. Y., & Manjunath, B. S. (1997). NeTra: A toollative box for navigating large image databases. In *Proceedings of IEEE Conference on Image Processing* (pp. 568–571).

Mohamed, A., Khellfi, F., Weng, Y., Jiang, J., & Ipson, S. (2009). An efficient image retrieval through DCT histogram quantization. In *Proceedings of International Conference on Cyber-Worlds* (pp. 237–240).

Nezamabadi-pour, H., & Kabir, E. (2004). Image retrieval using histograms of unicolour and bi-colour blocks and directional changes in intensity gradient. *Pattern Recognition Letters, 25,* 1547–1557.

Niblack, W., Barber, R., Equitz, W., Flickner, M., Glasman, E., Petkovic, D., et al. (1993). The QBIC project: Querying images by content using colour, texture and shape. *SPIE—The International Society for Optical Engineering, I Storage and Retrieval for Image and Video Databases, 1908,* 173–187.

Shih, J. L., & Chen, L. H. (2002). Colour image retrieval based on primitives of colour moments. In *Proceedings of IEEE Vision, Image and Signal Processing* (pp. 88–94).

Shim S.-O., & Choi, T.-S. (2003), Image indexing by modified color co-occurrence matrix, In *Proceedings of International Conference on Image Processing, 3:III* 2493–436

Smith, J. R., & Chang, S.-F. (1996). VisualSEEk: A fully automated content-based image query system. In *ACM Multimedia* (pp. 87–98).

Stricker, M. A., & Orengo, M. (1995). Similarity of colour images. *SPIE, 2420,* 381–392.

Swain, M. J., & Ballard, D. H. (1991). Colour indexing. *Computer Vision, 7,* 11–32.

Vadivel, A., Majumdar, A. K., & Shamik, S. (2003). Perceptually smooth histogram generation from the HSV colour space for content based image retrieval. In *Proceedings of International Conference on Advances in Pattern Recognition* (pp 248–251). Kolkata.

Vadivel, A., Sural, S., & Majumdar, A. K. (2008). Robust histogram generation from the HSV space based on visual colour perception. *International Journal of Signal and Imaging Systems Engineering, InderScience, 1*(3/4), 245–254.

Wang, X. (2009). A novel circular ring histogram for content-based image retrieval. In *Proceedings of 1st International Workshop on Education Technology and Computer Science* (Vol. 2, pp. 785–788).

Wang, S., & Qin, H. (2009). A study of order-based block colour feature image retrieval compared with cumulative colour histogram method. In *Proceedings of Sixth International Conference on Fuzzy Systems and Knowledge Discovery* (Vol. 1, pp. 81–84).

Chapter 5
Cluster Indexing and GR Encoding with Similarity Measure for CBIR Applications

5.1 Introduction

Digital media technology has seen enormous growth in recent times. The technology enables the users to store huge and voluminous data. As a result, new applications are deployed in many applications, especially in multimedia domain, say (Chuang et al. 2014), spatial information systems (Philbin et al. 2007), medical imaging (Wu et al. 2015), time-series analysis (Gouiffès et al. 2013), image retrieval systems (Wang and Chen et al. 2012a, Wang and Zeng et al. 2012b), storage and compression (Wang and Chen et al. 2009). These applications allowed the developers and users to store and retrieve information very easily. The availability of high-speed Internet bandwidth good access rate has also complemented real-time processing access. In addition to CBIR, text-based image retrieval (TBIR) is also used for information retrieval, but the query is in the form of text keywords (Vadivel and Shaila 2012). However, there are certain limitations and are handled by CBIR approach (Smeulders 2000). There has been a lot of research in the area of CBIR in academia and industry. In CBIR applications, the low-level features are used for representing colour, shape texture, etc. These features are high in dimension and suitably stored in a feature database with indexing scheme. During retrieval, the feature of query image is compared with all the database images. It is understood from most of the research that the colour content of images is very important. In addition, the human visual perception also understands the colour content (Vadivel et al. 2008). The image has high visual content and is often colour histograms (Jain and Vailaya 1996), colour layout (Smith and Chang 1996), etc.

The storage requirement, access time, transmission delay are some of the important issues to process the images data in WWW. Since the information is accessed and processed in real-time, may influence the performance of the retrieval system. The time to compare images in terms of their feature takes a lot of time by consuming large bandwidth. Also, the search time is linearly proportional to the size of the feature database. As a result, a suitable indexing mechanism is required as it

© Springer Nature Singapore Pte Ltd. 2018
S. G. Shaila and A. Vadivel, *Textual and Visual Information Retrieval using Query Refinement and Pattern Analysis*, https://doi.org/10.1007/978-981-13-2559-5_5

is an important concept in computer science (Heikkil and Pietikainen 2006), image analysis (Zhao and Pietikainen 2007), (Pietikainen et al. 2000), pattern recognition (Rahman et al. 2013; Liu et al. 2008). All these methods consider search time and precision of retrieval as challenging, especially in CBIR systems. In addition, the search process and time can be minimized by an effective feature extraction and representation scheme (Peitgen et al. 1992).

Various multimedia applications store the features and images in an unstructured way. The dimension of most of the features is large and uses different approaches to access the feature database such as clustering, indexing. This has caused the curse of dimensionality issue, and thus, the retrieval performance is degraded. As a result, effective methods are required to reduce the dimension of feature. Without any distortion to original meaning. Torres and Kunt (1996) have proposed encoding scheme to get rid of irrelevant bin values and thus exploited the redundancy. In recent times, features are encoded on individual cube and thus miss the redundancy of group of image features. The redundancy can be exploited by understanding the group in terms of colour, texture, shape, etc. The redundancy between the groups can improve the bit rate. Since the feature is grouped and interrelationship is maintained, a suitable similarity measure is also required. A fast similarity measure is required for various real-time applications, say machine learning (Pappis and Karacapilidis 1993), document retrieval (Recupero 2007), data compression (Cilibrasi and Vitányi 2005), bioinformatics (Lord et al. 2003), (Yu and He 2011) and data analysis (Le and Ho 2005).

This chapter addresses all the above-mentioned issues. An indexing scheme and encoding scheme with a similarity measure are proposed. The feature dimension is used for indexing. Golomb–Rice (GR) coding is applied for encoding the feature values, and a suitable similarity measure is proposed. The rest of the chapter is organized as follows. In Sect. 5.2, the related work is reviewed and Sects. 5.3, 5.4, 5.5, 5.6 and 5.7 explain the presented approach in detail. Section 5.8 presents the experimental results and concludes the chapter in the last section.

5.2 Literature Review

This section presents the literature review in three subsections. The Sect. 5.2.1 presents the review on dimension reduction of feature database. The last subsection reviews the approaches on similarity measures.

5.2.1 Review on Indexing Schemes

The indexing mechanism complements the retrieval applications by reducing the feature space and computation time. The GPU hardware improves the performance of any computing, and researchers from academia and industry have exploited the architecture parallel programming. The sequential indexing scheme is ineffective on GPU architecture, and these kinds of indexing structure are broadly classified into tree-based and hash-based structure. Tree-based indexing structure is divided into balanced and unbalanced structures. Knuth (1997) has proposed an unbalanced tree called binary tree having two children. The structure is degenerative and expands the search space. Some of the well-known balance tree structures in the memory indexing category are B trees (Bayer and McCreight 1972), B+ trees (Jannink 1995), red–black trees (Guibas and Sedgewick 1978), AVL trees (Knuth 1997) and T trees (Lehman and Carey 1986). T tree (Lee et al. 2007) and B-trees (Chen et al. 2002) are cache conscious structure and run SIMD instructions (Zhou and Ross 2002; Kim et al. 2010). All of these tree-based structures take more time to re-organize the tree structure and key comparison. As a result, there is a considerable amount of degradation in the performance (Bohm et al. 2011). In addition, the interleaved series of computation of the tree structure is found to be unsuitable for parallel computation (Bentley et al. 1975). The spectral hashing (SH) (Weiss et al. 2012) and local sensitive hashing (LSH) (Andoni and Indyk 2008) are structured based on hash. These structures handle the above issues, and however, these algorithms support only generic distance categories. A chained bucket hashing (Knuth 1997) is proposed, and no re-organization is required, since there is a fixed hash table size. However, most of the tree is degenerated in the form of linked list data structure, and as a result, tree is linear data structure.

Extendible hashing (Fagin et al. 1979), linear hashing (Litwin 1980) and modified linear hashing (Lehman and Carey 1986), are another kind of tree structures with dynamic reorganization capabilities. Ross (2007) has proposed a scheme to handle multiple hash buckets. A brute force scheme has been proposed on CUDA architecture (Garcia et al. 2008, 2010). The CUBLAS algorithm has been executed for retrieval applications (NVIDIA 2007). Cayton (2012) has proposed Random Ball Cover (RBC) on GPU platform. The random subsets are considered as reference for decomposing the space and has cower semantics cover for feature database. Korytkowski et al. (2015) have proposed a scheme to combine CBIR classification scheme with RDBMS. The conventional indexing schemes along with SQL queries are formulated to effectively perform retrieval task.

In addition to above, the indexing structure is further classified into data and space partitioning. The R tree and other similar tree are an example of data partitioning. The data space is bound with respect to regions, and it is extended hierarchically. The grid file, KDB tree and pyramid tree are examples under space partitioning category. However, all of these methods work well only when the dimension of the feature is low. One of the ways to handle this issue is to form clustering so that the dimension of the images feature is reduced for narrowing down the search space. Some of the known classification approaches are (Bezdek and Kuncheva 2001) using nearest neighbour algorithms (Muja and Lowe 2014) and hashing algorithms (Andoni and Indyk 2008), and all these methods have supported high-dimensional indexing to improve retrieval. Similarly, the partitioning approaches are classified into partitioning, hierarchical, grid-based and density-based algorithms. K-mean clustering is a partitioning algorithm (Chen et al. 2011). The CHAMELEON is an example of hierarchical algorithms (Karypis et al. 1999). Both of the above approaches require termination condition and are sensitive to the order. The hierarchical clustering strategies are classified into agglomerative and division approach. The images are grouped as high similarity images (Stehling et al. 2001). Bhatia (2005) has introduced hierarchical clustering approach and represented as hierarchical model. Kinoshenko et al. (2005) have proposed a scheme, where the images are partitioned into various sets. The query is split into sub-classes to find the similar images from database. Rakesh et al. (1998) have proposed to quantize the data space into different cells. The indexing and retrieval are continued to quantized cells. However, the number of quantized cell for each dimension is the key factor in the performance.

Kailing et al. (2004) have proposed SUBCLU to detect the clusters having different shapes and position. Some of the cluster-based techniques assign weight to the features with different dimensions. Yu and Zhang (2003) and Xu et al. (2009) have proposed cluster tree and cluster tree* for indexing the features with high dimension. The wavelet transformation is applied, and its coefficients measure the roughness and scale of implementation of indexing. Based on the issue of cluster tree*, a CBC tree is proposed (Xu et al. 2007), and however, this tree has failed in calculating different threshold values. Jouili and Tabbone (2012) have proposed a graph indexing method. The graph datasets are permitted to have intersecting clusters. As a result, one graph is belonging to multiple clusters. The graph database is viewed as more number of classes by using the median value of graph. However, the intersection between the clusters has reduced the performance drastically. The self-organizing map (SOM) is the type of neural network used to quantize and preserve the data topology for classification and retrieval (Kohonen 1997). The query data is compared with neurons in the neural network for retrieval. Laaksonen et al. (2002) have proposed PicSOM for retrieval, which apply TSOM. The feature is clustered from dense to sparse, i.e. from top to bottom, and the field structure of the tree is considered as one of the drawbacks.

Wang and Nie et al. (2013a, 2013b) have indexed the feature database using multi-modal spectral clustering and sparse multi-modal machine. The patterns of directional local extremes along with magnitude value are considered as patterns for CBIR applications (Vijaya Baskar Reddy and Rama Mohan Reddy 2014). However, this method takes more time to perform the search. In CBIR applications, it is often found that there is impreciseness in defining the query and feature database. Thus, it is important to manage this kind of properties in the database also. Some of the well-known databases are fuzzy database (Daoudi and Idrissi 2015) FOOD index (FI) (Yazici et al. 2008) and B+ tree-based indexing (2BPT) (Barranco et al. 2008). All these methods handle flexible queries and are found to be not suitable for multimedia retrieval applications. FOOD is yet another scheme for fuzzy approaches to handle object-oriented fuzzy database. This method requires that there has to be a linguistic variable for fuzzy restrictions. One of the drawbacks of these approaches is that it is expensive in terms of computational time and storage.

By considering all the research issues, an indexing scheme is presented to narrow down the search space to improve the retrieval performance.

5.2.2 *Literature Review on Encoding Approaches*

The feature of the image is represented as high-dimensional data. The content of the feature is considered as pivot for indexing the database. The retrieval time from these kinds of databases is on the higher side. Also, the huge storage space is required for storing the features. The storage space and retrieval time can be improved by reducing the size of the feature database. One of the effective ways is to reduce the dimension of the feature vector by retaining only the important information. Bronstein et al. (2006) have proposed a mechanism to encode the feature database effectively. Both Qiu (2003) and Poursistani et al. (2013) have proposed various encoding schemes to reduce the dimension of feature database. Qiu et al. (2003) have used block truncation coding (BTC). The image is encoded in RGB colour space using BTC algorithm. A bitmap of the image is generated along with couple of quantizes. Guo et al. (2013) have proposed a scheme and encoded image feature using ODBTC. Based on human visual perception, bitmap image is generated in the encoding stage (Guo and Wu 2009; Guo 2010). A multifactor co-relation is proposed to describe feature of images. The co-relation between R, G, B in the RGB colour space is extracted as feature (Wang and Wang 2014). Wang and Wang (2013) have proposed structure element descriptor to represent low-level features such as texture and colour.

The structure elements' descriptor (SED) constructs the structure elements' histogram (SEH) with 72 bins in HSV colour space. It is observed that while images are represented as tree, the encoding has shown improvement. The region-based image representation using BSP tree has decent performance improvement in terms of retrieval (Salembier and Garrido 2000; Cho et al. 2003). The BSP tree partitions the regions as small sub-regions and found that there is no meaning in sub-regions. The height of the tree is proportional to the objects and content has no clarity.

Kramm (2007a) has proposed a cluster-level encoding scheme. This method has failed to handle global redundancy. Kramm (2007b) has proposed PIF-based approach to encode similar parts in image. However, the decoding and encoding time are so high (Kramm 2008). Pappas (2013) has proposed perpetual model of coding, and the overlapping texture information is used for representing texture. These regions are connected, and corresponding texture map is derived. One of the drawbacks of this approach is the encoding time. Salembier and Garrido (2000) have analysed the relative performance between L778 and Lempel–Ziv–Welch (LZW) coding and has limitation in terms of storage and retrieval time (Zhen and Ren 2009). Barnsley and Sloan (1988) have proposed fractal block coding scheme and is extended by Jacquin (1993). The grey level is modified, and a suitable transformation is carried out (Wang et al. 2010). Genetic algorithm is applied to image compression and image retrieval. Wang and Wang (2008) have deployed block-based encoding. An adaptive plane strategy is used, and however, it considers only less sub-blocks only. The vector of aggregated local descriptors has been proposed for encoding. This approach uses local representation in terms of descriptors (Delhumeau et al. 2013; Jegou et al. 2010) and Fisher vector (FV) (Perronnin and Dance 2007; Perronnin et al. 2010). These descriptors are compressed using binary code (Perronnin et al. 2010) and product quantization (PQ) (Jegou et al. 2011). The search and retrieval are done in the compressed domain and reduce the space and time. In addition to all the above-mentioned approach, principal component analysis (PCA) (Lu et al. 2008), local linear embedding (LLE) (Roweis and Saul 2000) and locality preserving projection (LPP) (Yu et al. 2006) have also been used for reducing the redundancy. However, all these methods are unable to maintain descriptive ability and thus have achieved lower precision of retrieval.

Based on the above discussion, this chapter presents an encoding scheme with lesser number of bits to represent the feature.

5.2.3 Literature Review on Similarity Metrics

In computing paradigm, the similarity measure is very important to find the nearest neighbour. In CBIR applications, the similarity measure computes the distance between the query and database feature and the retrieval set is ranked based on the distance value. Well-known distance metrics are Manhattan (Vadivel et al. 2003), Euclidean (Vadivel et al. 2003), Mahalanobis (Vadivel et al. 2003; Swain and Ballard 1991). These approaches measure the distance of features having equal dimensions. While there is a difference between in the bin of query and data-based feature, the comparison is done between bin-to-bin. The index of the bins is used as reference to compute the distance. The earth mover's distance (EMD) (Rubner et al. 1997), Minkowski-form distance (Rubner et al. 1997), Histogram intersection (HI) (Rubner et al. 1997), etc., are well-known distance under this category. All these distance measures calculate the distance in feature space of their convenient.

The nearest neighbour scheme is categorized as partitioning tree, hashing scheme ad neighbouring graph. Bentley (1975) has introduced K–D tree, and it is suitable only for low-dimensional vectors.

Sproull (1991) have proposed multi-randomized K–D tree to improve approximate nearest neighbour search. Further, PCA tree (Sproull 1991, random projection (RP) trees (Dasgupta and Freund, 2008) and trinary projection trees (Jia et al. 2010) have also been proposed. All these algorithms have overhead to handle multiple dimensions, which is considered as a drawback. The hash information has been used to derive a similarity measure, and one such method is local sensitive hashing (LSH). Lv et al. (2007) have proposed multi-probe LSH. This has reduced the number of hash table and size of LSH forest (Bawa et al. 2005). These schemes are heavily dependent on hashing quality, which in return decides the performance. Some of improving hashing mechanisms are parameter sensitive hashing (Shakhnarovich et al. 2003), spectral hashing (Weiss et al. 2008), randomized LSH hashing from learned metrics (Jain et al. 2008), kernelized LSH (Kulis and Grauman 2009), learnt binary embeddings (Kulis and Darrell 2009), shift-invariant kernel hashing (Raginsky and Lazebnik 2009), semi-supervised hashing (Wang and Kumar et al. 2010), optimized kernel hashing (He et al. 2010) and complementary hashing techniques

(Xu et al. 2011), etc. A similarity measure is proposed using the concept of graphs. The vertices are points and edges connect the points to realize the nearest neighbour. The graph-based strategy is followed to achieve nearest neighbours. In (Sebastian and Kimia 2002), the BFS search is adapted and graph is explored from a seed point, which is well separated. Hajebi et al. (2011) have implemented BFS on K-NN graph point, and however, it is challenging to fix the initial point and handling hill climbing strategy. Wang et al. 2012a, b have constructed approximate neighbour graph by reducing the graph construction cost. However, this algorithm makes use of heuristic to handle the values to parameters manually. Fisher vector (FV) and vector of aggregated local descriptors (VLAD) are encoding scheme based on local descriptors for reducing the dimensions to facilitate fast comparison. Mojsilovic et al. (2000) have defined a distance based on human perception. Both subjective experiments and multidimensional aspect of human visual perception are used. The visual information of image and text are combined to have probability-based similarity measure (Barnard et al. 2003). The fuzzy information of regions is used for proposing a similarity measure (Chen and Wang 2002).

It is imperative from above discussion that most of the similarity measure has curse of dimensionality issue. This increases the storage space requirement and computational time. All these approaches fail, while the features are inter- and intragroup. In this chapter, BOSM is discussed and it handles the difference in the dimension of query and database features.

5.3 Architectural View of Indexing and Encoding with Similarity Measure

The approach contains three functional units. These units are named as indexer, encoder and similarity measure. Figure 5.1 shows the schematic of the approach. As initial step, the colour content of images is captured and resented as colour histogram. The dimension of these histograms is truncated, coded and indexed in the second stage. The distance between query and database feature is calculated in the last stage.

5.4 Histogram Dimension-Based Indexing Scheme

One of the important tasks of SBIR system is to retrieve relevant information. These systems rank the retrieval set based on rank, which is calculated by the distance metrics. Let T_{dist} and T_{rank} be time to calculate the distance and time to rank the retrieval, respectively. Most of the time value of T_{dist} and T_{rank} are high, and this is due to an exhaustive search of the feature database. This value can be reduced by suitable indexing scheme. In this chapter, the colour information is extracted and histogram is constructed as specified in Vadivel et al. (2008). The bin values are in

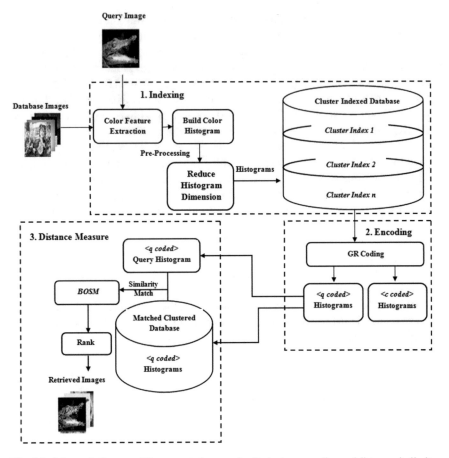

Fig. 5.1 Schematic diagram of the presented approach of indexing, encoding and distance similarity measure

decimal format, and bins with zero and negligible values are neglected. This enables us to reduce the dimension and thus the requirement of storage space. Also, there is a difference in structure of query and database feature. Figure 5.2 depicts the sample histogram. The dot represents the refined bins, and empty cell denotes the redundant bins. Eq. (5.1) shows the dimension range. Here, range of the index level of the histogram is considered as 4.

$$c_m = [1 + (\text{IF}) * (m - 1) \le R_m \le (\text{IF}) * m] \qquad (5.1)$$

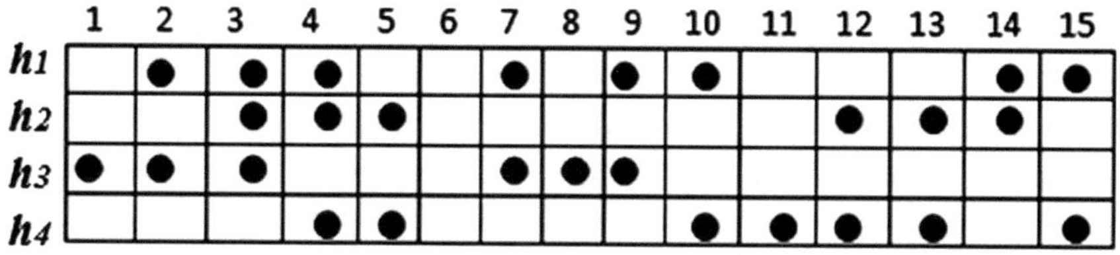

	1	2	3	4	5	6	7	8	9	10	11	12	13	14	15
h1		●		●		●			●		●	●		●	●
h2			●		●	●						●		●	●
h3	●	●		●					●		●	●			
h4				●		●					●		●	●	●	●

Fig. 5.2 Sample histogram with empty cells (bins)

Histogram Dimension	Dimension Range (R_m)	Cluster Index (C_m)	Index level (m)
64	49–64	C_4	4
48	33–48	C_3	3
32	17–32	C_2	2
16	01–16	C_1	1

Fig. 5.3 Sample indexing structure

In the above equation, c_m is cluster index and m represents level of index, which is in the range of 1–4. The R_m is dimension of histogram. In Eq. (5.2), the index factor is shown. Here, N is the dimension of histogram. A sample indexing structure is depicted in Fig. 5.3.

$$\text{IF} = \left(\frac{N}{M}\right), \text{ where } N = 64 \tag{5.2}$$

The index level (m) points to a cluster index (C_m) and the corresponding cluster the range of dimension as R_m. The value of IF and m is a sample in this chapter, and however, it can be fixed based on requirements, say size of the database. While performing query, the feature of query image is processed and truncated. Based on its dimension, the cluster index is calculated and query feature is compared only with the feature in the cluster. Below in Eq. (5.3), the total time to perform the search by this approach is represented and is found to be effective.

$$T_{\text{cluster Search}} = \{T_{\text{clu - match}} + T_{\text{dist}} + T_{\text{rank}}\} = \{C_{\text{Tclu - match}} + r\,T_{\text{dist}} + O(r \ \log r)\} \tag{5.3}$$

5.5   Coding Using Golomb–Rice Scheme

GR coding is found to be suitable for integer values. The bin value of the histogram is float and is converted into integer using a suitable multiplication factor (MF) as shown in Table 5.1. The GR coding is applied to these converted bin values, and finally, quotient (q) and reminder (r) part of the original histogram is obtained. GR coding has a tunable parameter, which decides the quotient and remainder parts of the histogram which is presented in Eq. (5.4) In this equation $N=0.69$, that is fixed based on the literature (Justin and Alistair 2006), n is the nonzero bins and B_i is the bin value (Fig. 5.4)

$$M = \sum_{i=1}^{n} \left(\frac{B_i}{n} \right) * W \tag{5.4}$$

Equations 5.5 and 5.6 divide the entire histograms into quotient and remainder part. Equations 5.5 and 5.6 calculate the quotient and remainder code of bin values. Based on the dimension of the feature, it is segregated into two parts, say quotient and remainder parts. Finally, the clusters are formed with these encoded features. Here, B_i represents the bin value of the histogram, and M is the tunable parameter of the respective cluster index.

$$\text{Quotient}(q) = \text{int}\left(\frac{B_i}{M} \right) \tag{5.5}$$

$$\text{Reminder}(r) = [B_i \bmod M] \tag{5.6}$$

Table 5.1 Sample value of multiplication factor

No. of zero after decimal point	MF
0.0(1)	100.0
0.00(2)	1000.0
0.000(3)	10000.0

Histogram Dimension	Dimension Range (R_m)	Cluster Index (C_m)	Index level (m)	M-Parameter
64	49–64	C_4	4	19
48	33–48	C_3	3	25
32	17–32	C_2	2	39
16	01–16	C_1	1	76

Fig. 5.4 Values of M for various indexing level

Table 5.2 Encoded feature (histogram)

Bin value of histogram (B_i)	Truncated Bin value	Multi-fact (MF)	Final MF	Trunc*(MF)	M	Q
0.027412	0.0274	100.0		27		1
0.002836	0.0028	1000.0		2		0
0.005121	0.0051	1000.0	1000.0	5	25	0
0.020674	0.0206	100.0		20		0
0.004128	0.0041	1000.0		4		0
0.121321	0.1213	10.0		121		4

Quotient code	Reminder	Reminder code	Combined code
10	2	0010	100010
0	2	0010	0010
0	5	0101	0101
0	20	11011	11011
0	4	0100	0100
11110	21	11100	1111011100

Further, two different parts of the histogram are coded in unary coding or binary code as mentioned in GR coding scheme. The final histogram is expressed as the linear combination quotient and remainder codes as given in Eq. (5.7).

$$[ccode] = [qcode][rcode] \qquad (5.7)$$

Table 5.2 shows a sample histogram bin values with coded value by following entire procedure discussed in this section. In addition, the relative importance of *qcode* and *rcode* is analysed using entropy of the respective histogram. It is found through various experimental results (Sect. 5.7.2) that *qcode* has more information compared to *rcode* which is shown in Eqs. (5.8) and (5.9)

$$E(c\ code) = E(\langle q\ code \rangle \langle r\ code \rangle)$$
$$= E(\langle q\ code \rangle \langle 0 \rangle)$$
$$E(c\ code) \cong E(\langle q\ code \rangle) \quad \text{Since} < r\ code > \text{is negligible} \qquad (5.8)$$
$$E(c\ code) \cong w * E(q\ code) \qquad (5.9)$$

5.6 Similarity Measure

The BOSM is presented in this section, which has good precision of retrieval. The feature database is clustered and indexed as different levels as shown in Fig. 5.5.

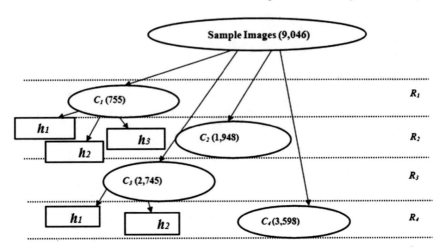

Fig. 5.5 View of database cluster

Table 5.3 Sample distance value

Overlap$_{(q)}$	Count_q	Overlap$_{(Degree)}$	h_q <q code>	h_k <q code>	XOR	Bindiff = $\sum$XOR
3	23	5/(23)	1110	110	1000	11111100
4			1110	111110	110000	
5			11110	11111110	11100000	
54			111110	110	111000	
55			1110	10	1100	

Here, $D = \{C_1, C_2, C_3, \ldots, C_m\}$ are database clusters. Here, C_m is the mth cluster and is represented as $C_m = \{h_1, h_2, h_3, \ldots, h_k\}$, where $h_1, h_2, h_3, \ldots, h_k$ represent histograms in the cluster. $B_i^{(hk)}$ is ith bin value of the histogram so that B_i ith is bin value h_k and B_j is jth *bin value of* h_q.

Given h_q and h_k, the BOSM computes the distance between them. The index which is common between query and databases histogram is considered, and the difference is calculated. The *count_q* and *count_k* are the number of nonzero bins in the query and database histogram, respectively. The overlap between query and database histogram is calculated using both of these values. The degree of overlap is calculated as follows

$$\text{Overlap}_{(Degree)} = \left(\frac{\text{Overlap}_q}{Count\_q}\right) \tag{5.10}$$

In above equation, the Overlap$_q$ is the number of bins overlapping and *Count_q* is the number of bins in the query. $B_i^{(hk)}$ and $B_j^{(hq)}$ are common bin values of query and database and will be calculated. The *Bindiff* is the difference between the bin values in binary and is given in Eq. (5.11), where B_j^{hk} is calculated when $i=j$ (Table 5.3).

Table 5.4 *BOSM* between histograms

C_m	Bindiff	Bindiff $_{(dist)}$	Normalized Bindiff $_{(dist)}$	Overlap$_{(Degree)}$	Normalized Overlap$_{(Degree)}$	BOSM
h_1	11111100	6	0.400	5/23 = 0.2173	0.1281	0.5281
h_2	1100	2	0.133	8/23 = 0.3478	0.2051	0.3381
h_3	1111000	4	0.266	11/23 = 0.4782	0.2820	0.548
h_4	111000	3	0.200	15/23 = 0.6521	0.3846	0.5846
		= 15		= 1.6954		

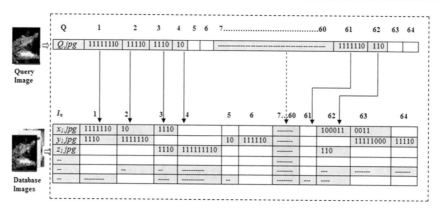

Fig. 5.6 Working principle of BOSM

$$Bindiff = Bindiff + \sum_{i=j=1}^{x} B_j^{hk} - B_i^{hq} \tag{5.11}$$

The difference between the bin values is calculated using Eq. (5.12), where the number of 1s present in the histogram is computed. The normalized value of Overlap$_{(Degree)}$ and *Bindiff* is added to get the final distance and is shown in Eq. (5.13).

$$Bindiff_{dist} = Bindiff \text{ (Number of 1s)} \tag{5.12}$$
$$BOSM = Normalized\left(Bindiff_{dist} + Overlap_{Degree}\right) \tag{5.13}$$

Table 5.4 presents BOSM value of different histograms.
Figure 5.6 depicts the working principle of BOSM.

5.6.1   *Algorithm—BOSM (H_q, H_k)*

Step 1: Consider Query Histogram h_q

Step 2: *Count_q* ← non-zero bin count of h_q

Step 3: Identify the matching cluster using *Count_q*

Step 4: Consider a sample histogram h_k of matched
cluster database

Step 5: *Count_k* ← non-zero bin count of h_k

//Compute *Overlap$_{(Degree)}$* and **Bindiff** between h_q and h_k //

Step 6: For Each bin index j of h_q
            For Each bin index i of h_k
                If (j==i) then
                    Overlap$_{(q)}$ = *Overlap$_{(q)}$* ++
                    Bindiff = Bindiff + ($B_j^{(hq)}$ − $B_i^{(hk)}$)
                Else If (i>j) then
                    Break the loop of h_k
            End loop
        End loop

Step7: *Overlap$_{(Degree)}$* = [*Overlap$_{(q)}$* /*Count_q*]

// BOSM //

Step 8: Repeat Step 4 to Step 7 for other histograms in
the matched cluster database

Step 9: Compute *Bindiff$_{(dist)}$* by counting number of 1's
present in *Bindiff*

Step 10: Normalize *Bindiff$_{(dist)}$* and *Overlap$_{(Degree)}$*

Step 11: Calculate *BOSM* between h_q and other histograms
of matched cluster database

Table 5.5 Details of benchmark datasets

Name of dataset	Categories	
	Count of images	Count of categories
Coral_10,000	10,000	100
MIT_9144	9,144	101
MIT_212	212	19
Caltech_101	9,146	101
Caltech_256	30,607	256

5.7 Experimental Results

This section of the chapter presents the performance evaluation of the proposed approaches. For experiments such as Coral, MIT and Caltech datasets are used. The number of images and their categories in each of these datasets are presented in Table 5.5.

The experimental results are presented in various subsections, and precision, recall and F1 score are used as metrics.

5.7.1 *Experimental Results on Coding*

In this section, the retrieval is presented using original and encoded histogram. The original histogram is represented as flat structure, which is not clustered and indexed. For the evaluation, the combined code ($ccode$) is considered. Initially, the experiment is conducted on Coral (Wang) dataset with 10,000 images. These 10,000 images are grouped into 100 images. The query set is randomly chosen from each category. Each image in the query set is presented as query, and retrieval set is obtained. While the images in the retrieval set match with the category of query images, it is marked as relevant. Similarly, for all the query images, the relevant images are found and the precision, recall and F1 score are calculated using Eq. (5.14)

$$\left. \begin{array}{l} \text{Precision} = \alpha/\beta \\ \text{Recall} = \alpha/\gamma \\ \text{F1-Score} = 2 \times \left(\frac{\text{Precision} \times \text{Recall}}{\text{Precision} + \text{Recall}} \right) \end{array} \right\} \qquad (5.14)$$

In Fig. 5.7, precision, recall and F1 score are depicted.

The precision and recall are calculated by analysing images having its rank within 50 in the retrieval set. The result is presented in Fig. 5.7 The difference in precision, recall and F1 score is almost zero for flat and encoded histogram. It is inferred from the result that encoded histogram can replace the flat structured original histogram. The perfromance of flat and cluster index histograms are evaluated on MIT dataset.

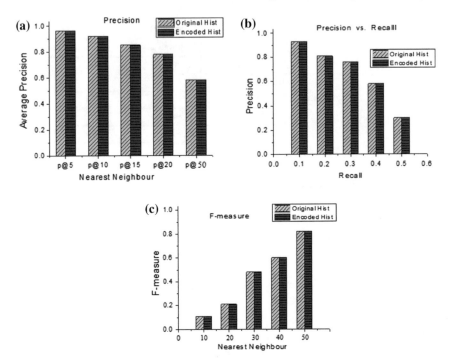

Fig. 5.7 Retrieval performance of encoded and flat histogram **a** precision, **b** precision versus recall, **c** F1 score

This has 9,356 images with object and texture dominant categories. The query set is prepared as mentioned earlier, and the relevant set is also found as mentioned in previous paragraph.

5.7.2 Retrieval Performance of BOSM on Various Datasets

In Fig. 5.8, the result is presented for *ccode* cluster *qcode*. Here, the *ccode* cluster is referred to histogram in encoded and clustered. The *qcode* is referred to encoded and clustered histogram. However, only the quotient code is used by the BOSM. Similar to above, the *ccode* flat and *qcode* flat are prepared and retrieved. Based on the results, the indexed histograms perform well compared to flat counterpart.

Figure 5.9 depicts a sample retrieval set from MIT dataset. The retrieval set in Fig. 5.9a, b is almost similar in visual content. However, there are irrelevant images in retrieval set presented in Fig. 5.9c, d. The reason is that the database is scattered, and it affects the precision of retrieval.

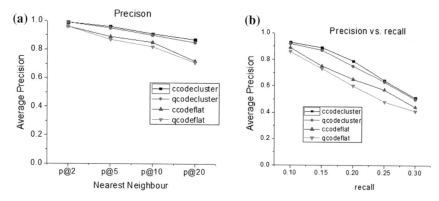

Fig. 5.8 Performance of BOSM on MIT dataset_212 **a** precision, **b** precision versus recall

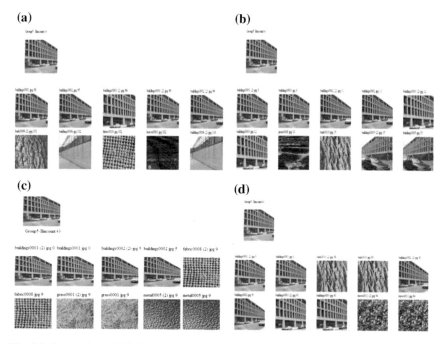

Fig. 5.9 Retrieval set MIT dataset_212 **a** clustered quotient code, **b** clustered combined, **c** flat *qcode*, **d** flat *ccode*

5.7.3 Evaluation of Retrieval Time and Bit Rate

Similar to the previous section, the flat and clustered structure of combined and quotient code histogram extracted from MIT dataset is used for calculating retrieval time. The retrieval time is calculated using Eq. (5.2) and is presented in Table 5.6. The bit rate is calculated using Eq. (5.14), and B_{qcode} and B_{ccode} are the number of

Table 5.6 Retrieval time

Code formats for indexing structures	Retrieval time (s)
Clustered quotient code	2.01
Flat quotient code	4.39
Clustered combined code	7.04
Flat combined code	10.41

Table 5.7 Bit ratio for MIT dataset

Cluster No.	Size of Cluster	B_{ccode}	B_{qcode}	B_R (%)
1	756	48,325	10,578	22
2	1,947	124,677	46,285	37
3	2,747	175,688	88,131	50
4	3,593	230,272	149,007	65

bits assigned for quotient and combined code histograms. Table 5.7 presents the bit rate for MIT datasets.

$$B_R = \left(\frac{B_{qcode}}{B_{ccode}} \right) * 100 \tag{5.14}$$

It is observed from Table 5.7 that the size of the quotient and combined code histogram decodes the dimension of the histogram. Based on all the results, quotient code has good retrieval performance, low retrieval time and good bit rate.

5.7.4   Comparative Performance Evaluation

The curse of dimensionality issue is evaluated in subsequent sections. The evaluation is consolidated further by comparing the results with some of the recent and relevant approaches.

5.7.4.1   Comparison Result

The performance of the proposed approach is compared with (Wang and Wang 2008, 2014, 2013; Kramm 2007a, b, 2008; Pappas 2013). The performance is evaluated on Caltech 101 and 256 datasets. The query is drawn from various classes of datasets, and the ground truth is calculated as discussed earlier. Tables 5.8 and 5.9 present the result with precision and F1 score.

Table 5.8 Performance comparison on Caltech dataset_101

Methods	Caltech dataset_101 Class ID									
	ID-1		ID-2		ID-3		ID-4		ID-5	
	P@10	F1	P@10	F1	P@10	F1	P@10	F1	P@10	F1
Presented approach	0.82	0.73	0.83	0.69	0.77	0.64	0.71	0.61	0.64	0.60
Kram (2007a)	0.68	0.59	0.64	0.612	0.56	0.51	0.70	0.52	0.50	0.46
Kram (2007b)	0.68	0.59	0.58	0.54	0.69	0.54	0.60	0.51	0.47	0.40
Kram (2008)	0.72	0.63	0.72	0.71	0.72	0.66	0.75	0.64	0.66	0.65
Wang (2008)	0.62	0.64	0.71	0.71	0.66	0.64	0.62	0.61	0.63	0.52
Wang (2013)	0.73	0.67	0.75	0.72	0.78	0.76	0.63	0.63	0.64	0.59
Pappas (2013)	0.57	0.53	0.73	0.67	0.73	0.69	0.60	0.52	0.70	0.63
Wang (2014)	0.79	0.76	0.79	0.71	0.79	0.66	0.71	0.70	0.65	0.62

Table 5.9 Performance comparison on Caltech dataset_256

Approaches	Caltech dataset_256 class ID									
	ID-1		ID-2		ID-3		ID-4		ID-5	
	P@10	F1	P@10	F1	P@10	F1	P@10	F1	P@10	F1
Presented approach	0.75	0.72	0.79	0.74	0.72	0.64	0.74	0.68	0.69	0.64
Kram (2007a)	0.66	0.56	0.62	0.53	0.52	0.49	0.66	0.56	0.44	0.46
Kram (2007b) (67)	0.64	0.63	0.54	0.51	0.66	0.54	0.56	0.54	0.44	0.41
Kram (2008)	0.69	0.64	0.73	0.71	0.64	0.62	0.75	0.67	0.63	0.57
Wang (2008)	0.68	0.72	0.68	0.63	0.58	0.54	0.64	0.63	0.59	0.56
Wang (2013) (63)	0.71	0.71	0.71	0.66	0.61	0.60	0.70	0.68	0.64	0.62
Pappas (2013)	0.52	0.51	0.68	0.65	0.72	0.64	0.58	0.50	0.66	0.61
Wang (2014)	0.76	0.74	0.78	0.74	0.68	0.64	0.72	0.71	0.68	0.62

Query Image

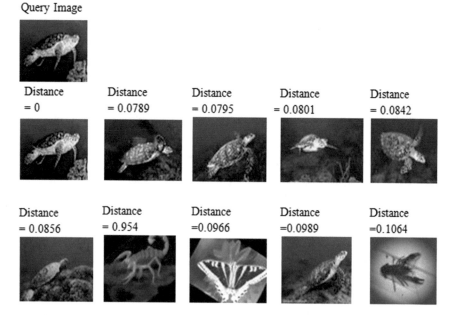

Fig. 5.10 Retrieval set from Caltech dataset _101

The sample retrieval set is depicted in Figs. 5.10 and 5.11. Given a query, the proposed encoding, indexing and similarity measure have retrieved most of the relevant images.

5.7.4.2 Performance of Similarity Measure

The proposed similarity measure is compared with earth mover's distance (EMD) (Rubner et al. 2000) and histogram intersection (HI) (Rubner et al. 1997). The working principle of these two algorithms is similar to that of the proposed approach. A query set with 200 images is drawn randomly from different classes of Caltech dataset, and result is presented in Fig. 5.12. It is found that the proposed similarity measure returns images in 3.28 s. In contrast, EMD and HI take 4.42 and 4.89 s, respectively. This trend is maintained for Caltech_256 dataset, where the proposed approach takes lesser time compared to the counterpart. Figures. 5.13 and 5.14 present another result, where the precision at different nearest neighbour is depicted. Another variant of the performance is presented in Tables 5.10 and 5.11.

Query Image

| Distance =0 | Distance = 0.0509 | Distance = 0.0515 | Distance = 0.0561 | Distance = 0.0598 |

| Distance = 0.0613 | Distance = 0.0689 | Distance = 0.0695 | Distance = 0.0701 | Distance = 0.0742 |

Fig. 5.11   Retrieval set from Caltech dataset_256

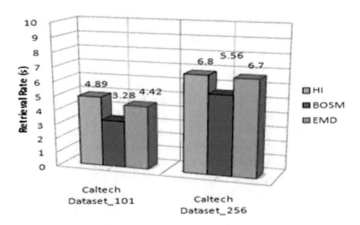

Fig. 5.12   Performance of similarity measure

It is observed from all the experimental results presented in this section that indexing scheme considerably reduced the retrieval time and storage space. The number of bits required to represent value f the histogram bin is also very low. The BOSM has handled the dimensionality issue effectively and achieved encouraging precision of retrieval. Overall, all the schemes proposed in this chapter are suitable for CBIR applications.

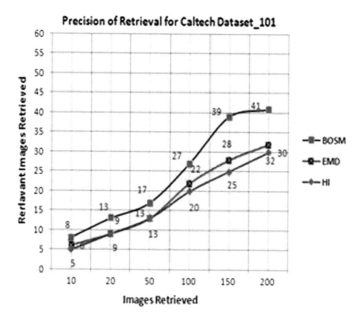

Fig. 5.13 Performances of distance measures for Caltech dataset_101

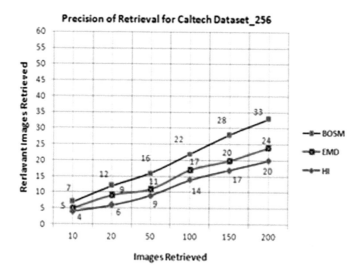

Fig. 5.14 Performance on Caltech dataset_256

Table 5.10 Precision on
Caltech dataset_101

Precision	Proposed (%)	EMD (%)	HI (%)
P@10	82	61	51
P@20	66	44	46
P@50	35	27	27
P@100	28	23	21
P@150	27	17.5	16.7
P@200	21	17	16.7

Table 5.11 Precision on
Caltech dataset_256

Precision	Proposed (%)	EMD (%)	HI (%)
P@10	81	61	51
P@20	64	46	46
P@50	33	27	27
P@100	28	23	21
P@150	27	18.8	16.7
P@200	20.4	15	15.8

5.8   Conclusion

In this chapter, three schemes are proposed for indexing, coding and calculating the distance between images. The aim of these schemes is to reduce the retrieval time, lower the storage space requirement and improve the precision of retrieval. The indexing mechanism has reduced the dimension of the feature by considering only relevant bins. Encoding scheme further reduced the size of the feature database by representing the bin values in quotient code of GR scheme. Further, the similarity measure has handled the variation in the size of query and database feature. The performance of all these schemes is evaluated on various well-known benchmark datasets. It is further consolidated by comparing the performance with some of the recently proposed counterpart schemes. Based on the retrieval results and analysis, all these three schemes perform well in CBIR applications.

References

Andoni, A., & Indyk, P. (2008). Near-optimal hashing algorithms for approximate nearest neighbor in high dimensions. *Communications of the ACM, 51*(1), 117–122.

Barnard, K., Duygulu, P., Forsyth, D., Freitas, N. D., Blei, D. M., & Jordan, M. I. (2003). Matching words and pictures. *Journal of Machine Learning Research, 3,* 1107–1135.

Barnsley, M., & Sloan, A. D. (1988). A better way to compress images. *BYTE, 13,* 215–233.

Barranco, C. D., Campana, J. R., & Medina, J. M. (2008). A b+-tree based indexing technique for fuzzy numerical data. *Fuzzy Sets System, 159,* 1431–1449.

Bawa, M., Condie, T., & Ganesan, P. (2005). LSH forest: Self-tuning indexes for similarity search. In *14th International Conference on World Wide Web*, pp. 651–660.

Bayer, R., & McCreight, E. M. (1972). Organization and maintenance of large ordered indices. *Acta Informatica, 1,* 173–189.

Bentley, J. L. (1975). Multidimensional binary search trees used for associative searcing. *Communications of the ACM, 18*(9), 509–517.

Bezdek, J. C., & Kuncheva, L. I. (2001). Nearest prototype classifier designs: An experimental study. *International Journal of Intelligent Systems, 16*(2), 1445–1473.

Bhatia, S. (2005). Hierarchical clustering for image databases. In *International Conference on Electro Information Technology*, pp. 6–12.

Böhm, M., Schlegel, B., Volk, P. B., Fischer, U., Habich, D., & Lehner, W. (2011). Efficient in-memory indexing with generalized prefix trees. In *der 14. GI-Fachtagung für Datenbanksysteme in Business, Technology und Web (BTW-2011)*.

Bronstein, A. M., Bronstein, M. M., & Kimmel, R. (2006). Generalized multidimensional scaling: A framework for isometry-invariant partial surface matching. *National Academy of Sciences, 103*(5), 1168–1172.

Cayton, L. (2012). Accelerating nearest neighbor search on many core systems. In *26th International Conference on Parallel & Distributed Processing Symposium (IPDPS '12)*, pp. 402–413.

Chen, L., Cui, B., & Lu, H. (2011). Constrained skyline query processing against distributed data sites. *IEEE Transaction on Knowledge and Data Engineering, 23*(2), 204–217.

Chen, S., Gibbons, P. B., Mowry, T. C., & Valentin, G. (2002). Fractal prefetching B_Trees: Optimizing both cache and disk performance. In *SIGMOD*, pp. 157–168.

Chen, Y., & Wang, J. Z. (2002). A region-based fuzzy feature matching approach to content-based image retrieval. *IEEE Transaction of Pattern Analysis and Machine Intelligence, 24*(9), 1252–1267.

Cho, S. Y., Chi, Z., Siu, W. C., & Tsoi, A. C. (2003). An improved algorithm for learning long-term dependency problem in adaptive processing of data structures. *IEEE Transaction of Neural Networks, 14,* 781–793.

Chuang, C.-H., Cheng, S.-C., Chang, C.-C., & Chen, Y.-P. P. (2014). Model-based approach to spatial–temporal sampling of video clips for video object detection by classification. *Journal of Visual Communication and Image Representation, 25,* 1018–1030.

Cilibrasi, R., & Vitányi, P. M. B. (2005). Clustering by compression. *IEEE Transaction of Information Theory, 51*(4), 1523–1545.

Daoudi, I., & Idrissi, K. (2015). A fast and efficient fuzzy approximation-based indexing for CBIR. *Multimedia Tools Applications, 74,* 4507–4533.

Dasgupta, S., & Freund, Y. (2008). Random projection trees and low dimensional manifolds. In: *40th Annual ACM Symposium on Theory of Computing* (pp. 537–546).

Delhumeau, J., Gosselin, P.-H., & Jegou, H. (2013). *Revisiting the VLAD image representation*. Spain: ACM Multimedia Barcelona.

Fagin, R., Nievergelt, J., Pippenger, N., & Strong, R. H. (1979). Extendible hashing—A fast access method for dynamic files. *ACM Transactions Database Systems, 4*(3), 315–344.

Garcia, V., Debreuve, E., & Barlaud, M. (2008). Fast k nearest neighbor search using GPU. In *IEEE Computer Society Conference on Computer Vision and Pattern Recognition Workshops (CVPR '08)*, Los Alamitos, CA, USA.

Garcia, V., Debreuve, E., Nielsen, F., & Barlaud, M. (2010). K-nearest neighbor search: Fast GPU-based implementations and application to high-dimensional feature matching. In *17th IEEE International Conference on Image Processing (ICIP '10)* Los Alamitos, CA, USA (pp. 3757–3760).

Gouiffès, M., Planes, B., & Jacquemin, C. (2013). HTRI: High time range imaging. *Journal of Visual Communication and Image Representation, 24,* 361–372.

Guibas, L. J., & Sedgewick, R. A. (1978). Dichromatic framework for balanced trees. In *FOCS*, 1978 (p. 8).

Guo, J. M. (2010). High efficiency ordered dither block truncation with dither array LUT and its scalable coding application. *Digital Signal Processing, 20*(1), 97–110.

Guo, J.-M., Prasetyo, H., & Su, H.-S. (2013). Image indexing using the colour and bit pattern feature fusion. *Journal of Visual Communication and Image Representation, 24,* 1360–1379.

Guo, J. M., & Wu, M. F. (2009). Improved block truncation coding based on the void-and cluster dithering approach. *IEEE Transaction of Image Processing, 18*(1), 211–213.

Hajebi, K., Abbasi-Yadkori, Y., Shahbazi, H., & Zhang, H. (2011). Fast approximate nearest-neighbor search with k-nearest neighbor graph. In *22nd International Joint Conference on Artificial Intelligence* (pp. 1312–1317).

He, J., Liu, W., & Chang, S. F. (2010). Scalable similarity search with optimized kernel hashing. In *International Conference on Knowledge Discovery Data Mining* (pp. 1129–1138).

Heikkil, M. A., & Pietikainen, M. (2006). A texture based method for modeling the back-ground and detecting moving objects. *IEEE Transaction of Pattern Analysis and Machine Intelligence, 28*(4), 657–662.

Jacquin, A. El. (1993). Fractal image coding: A review. *IEEE, 81*(10), 1451–1465.

Jain, P., Kulis, B., & Grauman, K. (2008). Fast image search for learned metrics. In *IEEE Conference on Computer Vision and Pattern Recognition* (pp. 1–8).

Jain, A., & Vailaya, A. (1996). Image retrieval using colour and shape. *Pattern Recognition, 29,* 1233–1244.

Jannink, J. (1995). Implementing deletion in B+-trees. *SIGMOD Record, 24*(1), 33–38.

Jegou, H., Douze, M., & Schmid, C. (2011). Product quantization for nearest neighbor search. *Transactions PAMI, 33*(1), 117–128.

Jegou, H., Douze, M., Schmid, C., & Perez, P. (2010). *Aggregating local descriptors into a compact image representation in CVPR.*

Jia, Y., Wang, J., Zeng, G., Zha, H., & Hua, X. S. (2010). Optimizing kd trees for scalable visual descriptor indexing. In *IEEE Conference on Computer Vision and Pattern Recognition* (pp. 3392–3399).

Jouili, S., & Tabbone, S. (2012). Hypergraph based image retrieval for graph based representation. *Journal of Pattern Recognition, 45*(11), 4054–4068.

Justin, Z., & Alistair, M. (2006). Inverted files for text search engines. *ACM Computing Surveys (CSUR), 38*(2).

Kailing, K., Kriegel, H.-P., & Kroger, P. (2004). Density-connected subspace clustering for high-dimensional data. In *International Conference on Data Mining (SIAM)* (pp. 246–257).

Karypis, G., Han, E.-H., & Kumar, V. (1999). CHAMELEON: A hierarchical clustering algorithm using dynamic modeling. *IEEE Computer, 32*(8), 68–75.

Kim, C., Chhugani, J., Satish, N., Sedlar, E., Nguyen, A. D., Kaldewey, T., et al. (2010). FAST: Fast architecture sensitive tree search on modern CPUs and GPUs. In *SIGMOD* (pp. 339–350).

Kinoshenko, D., Mashtalir, V., & Yegorova, E. (2005). Hierarchical partitions for content image retrieval from large-scale database. In *Machine Learning and Data Mining in Pattern Recognition.* Berlin: Springer.

Knuth, D. E. (1997). The art of computer programming. In *Fundamental algorithms* (3rd ed., Vol. 1). Addison-Wesley.

Kohonen, T. (1997). *Self-organizing maps.* Berlin, Heidelberg, New York: Springer.

Korytkowski, M., Scherer, R., Staszewski, P., & Woldan, P. (2015). Bag-of-features image indexing and classification in Microsoft SQL server relational database. In *IEEE 2nd International Conference on Cybernetics (CYBCONF)*, Gdynia, Poland (pp. 24–26).

Kramm, M. (2007a). Compression of image clusters using KarhunenLoeve transformations. In *International Conference on Electronic Imaging (SPIE), Human Vision XII* (Vol. 6492, pp. 101–106).

Kramm, M. (2007b). Image cluster compression using partitioned iterated function systems and efficient inter-image similarity features. In *IEEE Conference on Signal-Image Technologies and Internet-Based System* (pp. 989–996).

Kramm, M. (2008). Image group compression using texture databases. In *International Conference on Electronic Imaging (SPIE), Human Vision, XII* (Vol. 6806, p. 10).

Kulis, B., & Darrell, T. (2009). Learning to hash with binary reconstructive embeddings. In *23rd Advances in Neural Information Processing Systems* (Vol. 22, pp. 1042–1050).

Kulis, B., & Grauman, K. (2009). Kernelized locality-sensitive hashing for scalable image search. In *IEEE 12th International Conference on Computer Vision* (pp. 2130–2137).

Laaksonen, J., Koskela, M., & Oja, E. (2002). PicSOM-self-organizing image retrieval with mpeg-7 content descriptors. *IEEE Transaction Neural Networks, 13*(4), 841–853.

Le, S. Q., & Ho, T. B. (2005). An association-based dissimilarity measure for categorical data. *Pattern Recognition Letters, 26*(16), 2549–2557.

Lee, I. H., Shim, J., Lee, S. G., & Chun, J. (2007). CST-Trees: Cache sensitive T-Trees. In *DASFAA, LNCS 4443* (pp. 398–409).

Lehman, T. J., & Carey, M. J. (1986). A study of index structures for main memory database management systems. In *VLDB* (pp. 294–303).

Litwin, W. (1980). Linear hashing: A new tool for file and table addressing. In *VLDB*.

Liu, Y., Zhang, D., & Lu, G. (2008). Region-based image retrieval with high-level semantics using decision tree learning. *Pattern Recognition, 41*(8), 2554–2570.

Lord, P., Stevens, R., Brass, A., & Goble, C. (2003). Investigating semantic similarity measures across the Gene Ontology: The relationship between sequence and annotation. *Bioinformatics, 19,* 1275–1283.

Lu, H., Plataniotis, K. N., & Venetsanopoulos, A. N. (2008). Uncorrelated multilinear principal component analysis through successive variance maximization. In *25th International Conference on Machine Learning* (pp. 616–623).

Lv, Q., Josephson, W., Wang, Z., Charikar, M., & Li, K. (2007). Multiprobe LSH: Efficient indexing for high-dimensional similarity search. In *International Conference on Very Large Data Bases* (pp. 950–961).

Mojsilovic, A., Kovacevi, J., Hu, J., Safranek, R. J., & Ganapathy, S. K. (2000). Matching and retrieval based on the vocabulary and grammar of colour patterns. *IEEE Transaction on Image Processing, 9*(1), 38–54.

Muja, M., & Lowe, D. G. (2014). Scalable nearest neighbor algorithms for high dimensional data. *IEEE Transactions on Pattern Analysis and Machine Intelligence, 36.* https://doi.org/10.1109/tp ami.2014.2321376.

NVIDIA. (2007). CUDA CUBLAS library.

Pappas, T. N. (2013). The rough side of texture: texture analysis through the lens of HVEI. In *International Conference on Human Vision and Electronic Imaging (SPIE-IS&T), XVIII* (Vol. 8651, pp. 865110P-1–12).

Pappis, C. P., & Karacapilidis, N. I. (1993). A comparative assessment of measures of similarity of fuzzy values. *Fuzzy Sets and Systems, 56*(2), 171–174.

Peitgen, H.-O., Jurgens, H., Saupe, D. (1992). *Chao and fractals: New frontiers of science.* New York: Springer.

Perronnin, F., Anchez, J. S., & Mensink, T. (2010). Improving the Fisher kernel for large-scale image classification. In *ECCV*.

Perronnin, F., & Dance, C. R. (2007). Fisher kernels on visual vocabularies for image categorization. In *CVPR*.

Perronnin, F., Liu, Y., Sanchez, J., & Poirier, H. (2010). Large-scale image retrieval with compressed Fisher vectors. In *CVPR*.

Philbin, J., Chum, O., Isard, M., Sivic, J., & Zisserman, A. (2007). Object retrieval with large vocabularies and fast spatial matching. In *IEEE Conference on Computer Vision Pattern Recognition* (pp. 1–8).

Pietikainen, M., Ojala, T., Scruggs, T., Bowyer, K. W., Jin, C., & Hoffman, K. (2000). Overview of the face recognition using feature distributions. *Pattern Recognition, 33*(1), 43–52.

Poursistani, P., Nezamabadi-Pour, H., Moghadam, M. A., & Saeed, M. (2013). Image indexing and retrieval in JPEG compressed domain based on vector quantization. *Mathematical and Computer Modelling, 57*(5–6), 1005–1017.

Qiu, G. (2003). Colour image indexing using BTC. *IEEE Transaction of Image Processing, 12*(1), 93–101.

Raginsky, M., & Lazebnik, S. (2009). Locality-sensitive binary codes from shift-invariant kernels. In *Advances in Neural Information Processing Systems* (Vol. 22, pp. 1509–1517).

Rahman, S. A., Leung, M. K. H., & Cho, S.-Y. (2013). Human action recognition employing negative space features. *Journal of Visual Communication and Image Representation, 24*, 217–231.

Rakesh, A., Johanners, G., Dimitrios, G., & Prabhakar, R. (1998). Automatic subspace clustering of high dimensional data for data mining applications. In *International Conference Management of Data (ACM SIGMOD)* (pp. 94–105).

Recupero, D. R. (2007). A new unsupervised method for document clustering by using WordNet lexical and conceptual relations. *Information Retrieval, 10*(6), 563–579.

Ross, K. A. (2007). Efficient hash probes on modern processors. In *ICDE*.

Roweis, S. T., & Saul, L. K. (2000). Nonlinear dimensionality reduction by locally linear embedding. *Science, 290*(5500), 2323–2326.

Rubner, Y., Guibas, L. J., & Tomasi, C. (1997). The earth mover's distance, multi-dimensional scaling, and colour-based image retrieval. In *International Conference on Image Understanding Workshop (DARPA)* (pp. 661–668).

Rubner, Y., Tomasi, C., & Guibas, L. J. (2000). The earth mover's distance as a metric for image retrieval. *International Journal of Computer Vision, 40*(2), 99–121.

Salembier, P., & Garrido, L. (2000). Binary partition tree as an efficient representation for image processing, segmentation, and information retrieval. *IEEE Transaction of Image Processing, 9*(4), 561–576.

Sebastian, T. B., & Kimia, B. B. (2002). Metric-based shape retrieval in large databases. In *IEEE Conference on Computer Vision Pattern Recognition* (Vol. 3, pp. 291–296).

Shakhnarovich, G., Viola, P., & Darrell, T. (2003). Fast pose estimation with parameter-sensitive hashing. In *IEEE 9th International Conference on Computer Vision* (pp. 750–757).

Smeulders, A. W. M. (2000). Content-based image retrieval at the end of the early years. *IEEE Transaction of Pattern Analysis and Machine Intelligence, 22*, 1349–1379.

Smith, J. R., & Chang, S.-F. (1996). VisualSEEK: A fully automated content-based image query system. *ACM Multimedia*, 87–98.

Sproull, R. F. (1991). Refinements to nearest-neighbour searching in k dimensional trees. *Algorithmica, 6*(1), 579–589.

Stehling, R. O., Nascimento, M. A., & Falcao, A. X. (2001). An adaptive and efficient clustering based approach for CBIR in image databases. In *IDEAS* (pp. 356–365).

Swain, M. J., & Ballard, D. H. (1991). Colour indexing. *Journal of Computer Vision, 7*(1), 11–32.

Torres, L., & Kunt, M. (1996). *Video coding: The second generation approach*. Boston: Kluwer Academic Publishers.

Vadivel, A., Majumdar, A. K., & Sural, S. (2003). Performance comparison of distance metrics in content-based image retrieval applications. In *International Conference on Information Technology*, Bhubaneswar, India (pp. 159–164).

Vadivel, A., & Shaila, S. G. (2012). Smooth weighted approach for colour histogram construction using human colour perception for CBIR applications. *International Journal of Multimedia & its Applications, 4*(1), 113–125.

Vadivel, A., Shamik, S., & Majumdar, A. K. (2008). Robust histogram generation from the HSV space based on visual colour perception. *International Journal of Signal and Image System and Engineering, InderScience., 1*(3/4), 245–254.

Vijaya Bhaskar Reddy, P., & Rama Mohan Reddy, A. (2014). Content based image indexing and retrieval using directional local extrema and magnitude patterns. *International Journal of Electronic Communication (AEÜ), 68*, 637–643.

Wang, X.-Y., & Chen, Z.-F. (2009). A fast fractal coding in application of image retrieval. *Fractals, 17*(4), 441–450.

Wang, X.-Y., Chen, Z.-F., & Yun, J.-J. (2012a). An effective method for colour image retrieval based on texture. *Computer Standards & Interfaces, 34*(1), 31–35.

Wang, J., Zeng, G., Tu, Z., Gan, R., & Li, S. (2012b). Scalable k-NN graph construction for visual descriptors. In *IEEE Conference Computer Vision of Pattern Recognition* (pp. 1106–1113).

Wang, J., Kumar, S., & Chang, S. F. (2010). Semi-supervised hashing for scalable image retrieval. In *IEEE Conference on Computer Vision Pattern Recognition* (pp. 3424–3431).

Wang, X.-Y., Li, F.-P., & Wang, S.-G. (2009). Fractal image compression based on spatial correlation and hybrid genetic algorithm. *Journal of Visual Communication and Image Representation, 20*(8), 505–510.

Wang, H., Nie, F. P., & Huang, H. (2013a). Multi-view clustering and feature learning via structured sparsity. In *ICML* (pp. 352–360).

Wang, H., Nie, F. P., Huang, H., & Ding, C. (2013b). Heterogeneous visual features fusion via sparse multimodal machine. In *CVPR* (pp. 3097–3102).

Wang, X.-Y., & Wang, S.-G. (2008). An improved no-search fractal image coding method based on a modified gray-level transform. *Computers & Graphics, 32*(4), 445–450.

Wang, X. Y., & Wang, Z. Y. (2013). A novel method for image retrieval based on structure elements' descriptor. *Journal of Visual Communication and Image Representation, 24*(1), 63–74.

Wang, X. Y., & Wang, Z. Y. (2014). The method for image retrieval based on multi-factors correlation utilizing block truncation coding. *Pattern Recognition, 47*(10), 3293–3303.

Wang, X.-Y., Wang, Y.-X., & Yun, J.-J. (2010b). An improved no-search fractal image coding method based on a fitting plane. *Image and Vision Computing, 28*(8), 1303–1308.

Weiss, Y., Fergus, R., & Torralba, A. (2012). Multidimensional spectral hashing. In *12th European Conf. Computer Vision (ECCV '12)* (pp. 340–353). Springer, Berlin, Germany.

Weiss, Y., Torralba, A., & Fergus, R. (2008). Spectral hashing. In *Advances in Neural Information Processing System* (p. 6).

Wu, H.-T., Huang, J., & Shi, Y.-Q. (2015). A reversible data hiding method with contrast enhancement for medical Images. *Journal of Visual Communication and Image Representation, 31*, 146–153.

Xu, H., Wang, J., Li, Z., Zeng, G., Li, S., & Yu, N. (2011). Complementary hashing for approximate nearest neighbor search. In *IEEE International Conference on Computer Vision* (pp. 1631–1638).

Xu, H., Xu, D., & Lin, E. (2007). An applicable hierarchical clustering algorithm for content based image retrieval. In *International Conference on Computer Vision/Computer Graphics Collaboration Techniques and Applications (MIRAGE)* (pp. 82–92).

Xu, H., Xu, D., & Lin, E. (2009). An image index algorithm based on hierarchical clustering. In *5th International Conference on Intelligent Information Hiding and Multimedia Signal Processing* (pp. 459–462).

Yazici, A., Ince, C., & Koyuncu, M. (2008). Food index: A multidimensional index structure for similarity-based fuzzy object oriented database models. *IEEE Transaction Fuzzy System, 16*(4), 942–957.

Yu, G., & He, Q.-Y. (2011). Functional similarity analysis of human virus-encoded miRNAs. *Journal of Clinical Bioinformatics, 1*(1), 15.

Yu, W., Teng, X., & Liu, C. (2006). Face recognition using discriminant locality preserving projections. *Image and Vision Computing, 24*(3), 239–248.

Yu, D., & Zhang, A. (2003). ClusterTree: Integration of cluster representation and nearest-neighbour search for large data sets with high dimensions. *IEEE Transactions on Knowledge and Data Engineering, 15*(5), 1316–1337.

Zhao, G., & Pietikainen, M. (2007). Dynamic texture recognition using local binary patterns with an application to facial expressions. *IEEE Transaction of Pattern Analysis and Machine Intelligence, 29*(6), 915–928.

Zhen, C., & Ren, B. (2009). Design and realization of data compression in real-time database. In *10th International Conference on Computational Intelligence and Software Engineering* (pp. 1–4).

Zhou, J., & Ross, K. A. (2002). Implementing database operations using SIMD instructions. In *SIGMOD* (pp 145–156).

Notes

1. http://wang.ist.psu.edu/docs/related.
2. http://web.mit.edu/torralba/www/database.html.
3. http://www.vision.caltech.edu/Image_Datasets/Caltech101.
4. http://www.vision.caltech.edu/Image_Datasets/Caltech256.

Appendix

1. Smart information retrieval system. ftp://ftp.cs.cornell.edu/pub/smart/english. stop.
2. http://tartarus.org/∼martin/PorterStemmer/.
3. http://code.google.com/apis/soapsearch/reference.html.
4. http://www.iraqbodycount.org/database/.
5. http://www.trutv.com/library/crime/.
6. http://dl.dropbox.com/u/84084/GOV2.txt.
7. http://dl.dropbox.com/u/84084/clueweb09.txt.
8. http://dl.dropbox.com/u/84084/WT10g.txt.
9. http://www.99acres.com.
10. http://www.makemytrip.com.
11. http://www.indiaproperty.com.

© Springer Nature Singapore Pte Ltd. 2018
S. G. Shaila and A. Vadivel, *Textual and Visual Information Retrieval using Query Refinement and Pattern Analysis*, https://doi.org/10.1007/978-981-13-2559-5

Printed in the United States
By Bookmasters